A Hybrid Imagination

Science and Technology in Cultural Perspective

Synthesis Lectures on Engineers, Technology, and Society

Editor
Caroline Baillie, *University of Western Australia*

The mission of this lecture series is to foster an understanding for engineers and scientists on the inclusive nature of their profession. The creation and proliferation of technologies needs to be inclusive as it has effects on all of humankind, regardless of national boundaries, socio-economic status, gender, race and ethnicity, or creed. The lectures will combine expertise in sociology, political economics, philosophy of science, history, engineering, engineering education, participatory research, development studies, sustainability, psychotherapy, policy studies, and epistemology. The lectures will be relevant to all engineers practicing in all parts of the world. Although written for practicing engineers and human resource trainers, it is expected that engineering, science and social science faculty in universities will find these publications an invaluable resource for students in the classroom and for further research. The goal of the series is to provide a platform for the publication of important and sometimes controversial lectures which will encourage discussion, reflection and further understanding.

The series editor will invite authors and encourage experts to recommend authors to write on a wide array of topics, focusing on the cause and effect relationships between engineers and technology, technologies and society and of society on technology and engineers. Topics will include, but are not limited to the following general areas; History of Engineering, Politics and the Engineer, Economics , Social Issues and Ethics, Women in Engineering, Creativity and Innovation, Knowledge Networks, Styles of Organization, Environmental Issues, Appropriate Technology

A Hybrid Imagination: Science and Technology in Cultural Perspective
Andrew Jamison, Steen Hyldgaard Christensen, and Lars Botin
2011

A Philosophy of Technology: From Technical Artefacts to Sociotechnical Systems
Pieter Vermaas, Peter Kroes, Ibo van de Poel, Maarten Franssen, and Wybo Houkes
2011

Tragedy in the Gulf: A Call for a New Engineering Ethic
George D. Catalano
2011

Humanitarian Engineering
Carl Mitcham and David Munoz
2010

Engineering and Sustainable Community Development
Juan Lucena, Jen Schneider, and Jon A. Leydens
2010

Needs and Feasibility: A Guide for Engineers in Community Projects — The Case of Waste for Life
Caroline Baillie, Eric Feinblatt, Thimothy Thamae, and Emily Berrington
2010

Engineering and Society: Working Towards Social Justice, Part I: Engineering and Society
Caroline Baillie and George Catalano
2009

Engineering and Society: Working Towards Social Justice, Part II: Decisions in the 21st Century
George Catalano and Caroline Baillie
2009

Engineering and Society: Working Towards Social Justice, Part III: Windows on Society
Caroline Baillie and George Catalano
2009

Engineering: Women and Leadership
Corri Zoli, Shobha Bhatia, Valerie Davidson, and Kelly Rusch
2008

Bridging the Gap Between Engineering and the Global World: A Case Study of the Coconut (Coir) Fiber Industry in Kerala, India
Shobha K. Bhatia and Jennifer L. Smith
2008

Engineering and Social Justice
Donna Riley
2008

Engineering, Poverty, and the Earth
George D. Catalano
2007

Engineers within a Local and Global Society
Caroline Baillie
2006

Globalization, Engineering, and Creativity
John Reader
2006

Engineering Ethics: Peace, Justice, and the Earth
George D. Catalano
2006

A Hybrid Imagination: Science and Technology in Cultural Perspective

Andrew Jamison, Steen Hyldgaard Christensen, and Lars Botin

www.morganclaypool.com

ISBN: 9781608457373   paperback
ISBN: 9781608457380   ebook

DOI 10.2200/S00351ED1V01Y201104ETS016

A Publication in the Morgan & Claypool Publishers series
SYNTHESIS LECTURES ON ENGINEERS, TECHNOLOGY, AND SOCIETY

Lecture #12
Series Editor: Caroline Baillie, *University of Western Australia*
Series ISSN
Synthesis Lectures on Engineers, Technology, and Society
Print 1933-3633   Electronic 1933-3641

A Hybrid Imagination

Science and Technology in Cultural Perspective

Andrew Jamison
Aalborg University

Steen Hyldgaard Christensen
Aarhus University

Lars Botin
Aalborg University

SYNTHESIS LECTURES ON ENGINEERS, TECHNOLOGY, AND SOCIETY #12

MORGAN & CLAYPOOL PUBLISHERS

ABSTRACT

This book presents a cultural perspective on scientific and technological development. As opposed to the "story-lines" of economic innovation and social construction that tend to dominate both the popular and scholarly literature on science, technology and society (or STS), the authors offer an alternative approach, devoting special attention to the role played by social and cultural movements in the making of science and technology. They show how social and cultural movements, from the Renaissance of the late 15th century to the environmental and global justice movements of our time, have provided contexts, or sites, for mixing scientific knowledge and technical skills from different fields and social domains into new combinations, thus fostering what the authors term a "hybrid imagination." Such a hybrid imagination is especially important today, as a way to counter the competitive and commercial "hubris" that is so much taken for granted in contemporary science and engineering discourses and practices with a sense of cooperation and social responsibility. The book portrays the history of science and technology as an underlying tension between hubris – literally the ambition to "play god" on the part of many a scientist and engineer and neglect the consequences – and a hybrid imagination, connecting scientific "facts" and technological "artifacts" with cultural understanding. The book concludes with chapters on the recent transformations in the modes of scientific and technological production since the Second World War and the contending approaches to "greening" science and technology in relation to the global quest for sustainable development. The book is based on a series of lectures that were given by Andrew Jamison at the Technical University of Denmark in 2010 and draws on the authors' many years of experience in teaching non-technical, or contextual knowledge, to science and engineering students. The book has been written as part of the Program of Research on Opportunities and Challenges in Engineering Education in Denmark (PROCEED) supported by the Danish Strategic Research Council from 2010 to 2013.

KEYWORDS

history of science and technology, hybrids, hybrid imagination, hubris, habitus, social movements, appropriate technology, contextual knowledge, cognitive praxis, romanticism, industrialization, environmentalism, modernism, modernization, globalization

Contents

Preface

I have spent most of my working life as a humanist among scientists and engineers, teaching non-technical material to science and engineering students. I started in the early 1970s with a course on science and society for natural science students at the University of Copenhagen, and then spent ten years at the University of Lund, directing educational activities at both the undergraduate and postgraduate levels in science and technology policy. In 1996, I returned to Denmark to teach science and engineering students at Aalborg University, where I have tried to integrate non-technical material - or "contextual knowledge" as we have come to call it - into science and engineering educational programs.

All the teaching I have done through the years is based on a conviction that an understanding of the cultural and social contexts of science and technology should be an integral part of science and engineering education. I have always felt that students will be better able to use their scientific and technical knowledge for beneficial purposes if they have learned something about how science and technology are used – and misused - in the broader society. My contention is that a hybrid imagination – mixing scientific and/or engineering competence with cultural awareness - has been central to many, if not most of the positive achievements of science and technology in human history. On the other hand, when scientists and engineers have not had sufficient understanding of the implications of what they were doing, they have tended to cause as many problems as they have solved, if not more.

As the years have passed, it has often seemed strange to me that there weren't more people doing this kind of teaching. For during the past three or four decades, as science and technology have ever more come to dominate our lives and permeate our societies, the existence of both humanity, as well as the non-humans with whom we share the planet, has come to be increasingly filled with risks and dangers. From climate change to the proliferation of weapons of mass destruction, from traffic congestion to the invasion of our privacy, science and technology are directly involved in many, if not most, of the central issues of the contemporary world.

In light of these developments, it is important that the relations between science, technology and society should be given serious attention in the education of scientists and engineers - as well as in the wider realms of public education and debate. But instead of providing opportunities and spaces for learning about these relations, far too many educators and educational officials have joined in the efforts to promote more science and technology, rather than contribute to a contextual understanding of the science and technology we already have. There is far too little attention devoted to contextual knowledge in most science and engineering educational programs, and in educational programs in the social sciences and humanities, science and technology tend to be conspicuous by their absence.

Even more disheartening, at least to me, is that most of the research that is carried out about science, technology and society (or STS) has little influence on the actual education of scientists and engineers, in large measure due to the gap that has emerged between research and education at most universities. At best, contextual knowledge is "added on" to the core curriculum in science and engineering programs in an instrumental way, providing an extra skill or two – in entrepreneurship, innovation, marketing, communication, or ethics – that is considered helpful for would-be scientists and engineers to be able to function more effectively in the global marketplace. Most often, however, it is simply ignored.

This book is based on a course that I gave in the spring term 2010 at the Technical University of Denmark. I would like to thank the students who made the course at DTU such a memorable experience that I decided to turn it into a book, and Ulrik Jørgensen, my partner in a new Danish research program on engineering education, who gave me the opportunity to teach the course. It has been written while Ulrik and I have been initiating PROCEED (Program of Research on Opportunities and Challenges in Engineering Education in Denmark), and I have tried to write it in such a way that it can provide a frame of reference and, more specifically, a set of concepts and specific examples for the research we are undertaking. A description of the program can be found on my website.

I would like to thank Mikael Hård, professor at Darmstadt University of Technology, and co-author of *Hubris and Hybrids. A Cultural History of Technology and Science* (Routledge 2005), which served as a textbook for the course for his earlier collaboration, and for asking me to write an article for a German journal that was the starting-point for this book. I would also like to thank Jette Holgaard, my colleague in Aalborg, for her help in developing the understanding of contextual knowledge that is developed here and, in particular, the typology that is presented in chapter two.

Ulrik Jørgensen, Niels Mejlgaard and Matthias Heymann, partners in PROCEED, and Carl Mitcham, Peter Kroes, and Caroline Baillie, international advisers for PROCEED, gave me some very helpful comments on early drafts of the manuscript. A special thanks to Caroline, as editor of this series, and Joel Claypool, its publisher, for their support and encouragement throughout the process of turning those early drafts into a finished book. I would also like to thank the Danish Strategic Research Council for its financial support. In the later phases of writing, my co-authors, Steen Hyldgaard Christensen, who initiated PROCEED, and Lars Botin, a colleague in Aalborg, have contributed material to chapters three, four, and five, and I would like to thank them for thus making the book an example of the "cross-fertilizing" approach to research that we are carrying out in PROCEED.

In order to keep the book relatively short and readable, I have written it in essay form, trying to follow the structure and style of the lectures that I presented, and using some of the powerpoint slides that I showed in the course to help lighten up the text. Since the book draws on many of my earlier writings and, in particular, on *Hubris and Hybrids*, I recommend curious readers who want to know more about particular topics to visit my previous books and articles – as well as the many sources in the reference list. I would be most pleased to offer suggestions for further reading or simply discuss the book with anyone who might be interested in getting a conversation going. The whole point is for this book to be used by both engineering teachers and students.

Andrew Jamison
April 2011

Acknowledgments

Portions of this book have appeared in substantially different form in the following:

The Story-Lines of Technological Change: Innovation, Construction and Appropriation, with Mikael Hård, in *Technology Analysis & Strategic Management*, volume 15, nr 1, 2003

Hubris and Hybrids: A Cultural History of Technology and Science, co-author Mikael Hård, New York, Routledge, 2005

On Nanotechnology and Society, in *The Age of Nanotechnology*, edited by Nirmala Rao Khadpekar, Hyderabad, Icfai Books, 2007

To Foster a Hybrid Imagination: Science and the Humanities in a Commercial Age, in *NTM – Zeitschrift für Geschichte der Wissenschaften, Technik und Medizin*, nr 1, 2008

Can Nanotechnology Be Just? On Science, Technology and the Emerging Movement for Global Justice, in *Nanoethics*, 2009, volume 3, nr 2

The Historiography of Engineering Contexts, in *Engineering in Context*, edited by Steen Hyldgaard Christensen, Bernard Delahousse, and Martin Meganck, Aarhus, Academic, 2009

Contextualizing Nanotechnology Education: Fostering a Hybrid Imagination in Aalborg, Denmark, with Niels Mejlgaard, in *Science as Culture*, volume 19, nr 3, 2010

Climate Change Knowledge and Social Movement Theory, in *Wiley Interdisciplinary Reviews: Climate Change*, November 2010

Ecology and the Environmental Movement, in *Ecology Revisited: Reflecting on Concepts, Advancing Science*, edited by Astrid Schwarz and Kurt Jax, Springer, 2011

Andrew Jamison
April 2011

CHAPTER 1

Introduction

Imagination is not to be divorced from the facts: it is a way of illuminating the facts. It works by eliciting the general principles which apply to the facts, as they exist, and then by an intellectual survey of alternative possibilities which are consistent with those principles. It enables men to construct an intellectual vision of a new world, and it preserves the zest of life by the suggestion of satisfying purposes. ...The tragedy of the world is that those who are imaginative have but slight experience, and those who are experienced have feeble imaginations. Fools act on imagination without knowledge; pedants act on knowledge without imagination. The task of a university is to weld together imagination and experience.

Alfred North Whitehead, *The Aims of Education* (1929)

hy′brid (hī′brid) n. 1. the offspring of animals or plants of different varieties, species, or genera; a cross-breed or mongrel. 2. any product or mixture of two heterogeneous things.

The Webster Handy College Dictionary (1956)

1.1   AN AGE OF HYBRIDITY

As John Reader and Caroline Baillie have discussed in previous books in this series, globalization raises a number of basic challenges for engineering and engineering education [Baillie, C., 2009, Reader, J., 2006]. If they are to be able to carry out their work in a fair and responsible manner in a globalized world, engineers need to know a good deal more than merely how to make technical artifacts and technological systems function effectively. They need to be able to combine their technical and scientific knowledge with an understanding of how the wider world operates. In particular, as Downey, G. [2010] has recently argued, the globally-competent engineer needs to be educated about the very different cultures and cultural values that affect engineering work. As such, an important task of science and engineering education is to foster what we will be characterizing in this book as a hybrid imagination, by which we mean the mixing of scientific and technical knowledge with cultural awareness.

Globalization breeds hybrids. With so many people on the move, in search of a better life somewhere else, hybridity has become a sign of the times. More and more people in the world today form what cultural theorists call "hybrid identities" as a way to express themselves and understand who they are [Bhabha, H., 1994]. In a good deal of contemporary art, music and literature, previously separated genres, languages, styles and traditions are remixed in new kinds of hybrid combinations, as writers and artists reflect on their experiences through their imaginations. Salman Rushdie, whose book, *The Satanic Verses*, forced him into hiding to escape the Iranian president Khomeini's death threat in 1989, has put it this way:

> *The Satanic Verses* celebrates hybridity, impurity, intermingling, the transformation of new and unexpected combinations of human beings, cultures, ideas, politics, movies, songs. It rejoices in mongrelization and fears the absolutism of the Pure. *Mélange*, hotchpotch, a bit of this and a bit of that is *how newness enters the world*, and I have tried to embrace it. *The Satanic Verses* is for change-by-fusion, change-by-conjoining. It is a love-song to our mongrel selves [Rushdie, S., 1992, p. 394].

Hybridity is also increasingly important in the making of contemporary science and technology. On the one hand, there is a combination of what has historically been characterized as science and what has historically been characterized as technology into an array of "technosciences" that transcend traditional academic disciplines and professional identities. On the other hand, there is an ever more intimate connecting and interacting between science, technology and the broader society, as the relative autonomy of universities and the "academic freedom" that was once considered so important for scientific and technological development has increasingly been replaced by "academic capitalism"[Shapin, S., 2008, Slaughter, S. and Rhoades, G., 2004].

Among those who study the relations between science, technology and society (STS), hybridity and hybridization have become popular terms. Haraway, D. [1991] has written a manifesto for what she terms cyborgs: "theorized and fabricated hybrids of machine and organism." Latour, B. [1993] has characterized contemporary reality in terms of a "proliferation of hybrids" between humans and non-humans and calls for the overthrow of the "modern constitution" that was established in the 17th century separating the study of nature from the study of society. Society and humanity are no longer distinguishable from non-human things, according to Latour, and therefore science and technology need to be reorganized in ways that take this postmodern - or what he prefers to call "nonmodern" - condition into account.

Hybridization is also seen as a central ingredient in the new "transdisciplinary" mode of knowledge production, or so-called "mode 2," that has been described by Michael Gibbons and his co-authors in *The New Production of Knowledge* [Gibbons et al., 1994]: "Hybridization reflects the need of different communities to speak in more than one language in order to communicate at the boundaries and in the spaces between systems and subsystems"[Gibbons et al., 1994, p. 37].

In such fields as genetic engineering, information technology, sustainability science, media engineering and nanotechnology, as well as in many older fields, scientists and engineers are compelled, both for reputational and funding purposes, to bring together skills and knowledge from different areas of science and engineering, as well as from different parts of society. They need to transgress disciplinary boundaries and combine non-technical skills with their scientific and technical competence in order to produce knowledge in what Gibbons and his co-authors have termed "contexts of application."

1.2   A HYBRID IMAGINATION

The notion of a hybrid imagination that we will be presenting in this book builds on these ideas. More specifically, it is based on a cultural historical perspective in which a hybrid imagination is seen as a critical counterpoint to the "hubris" that has been fundamental throughout history in regard to scientific and technological development [Hård, M. and Jamison, A., 2005]. Hubris is a word that comes to us from the ancient Greeks; in the *Encyclopedia Britannica*, it is defined as

> overweening presumption suggesting an impious disregard of the limits governing men's actions in an orderly universe. It is the sin to which the great and gifted are most susceptible, and in Greek tragedy it is usually the basic flaw of the tragic hero (*Encyclopedia Britannica* 1979: V, 179).

In his culturally critical writings of the 1960s and 1970s, the Finnish philosopher Georg Henrik von Wright applied the term to science and technology, characterizing the "particular hubris of the modern technological way of life" with an "unreasonable redirection of nature's causality for human purposes"[von Wright, G.H., 1978, p. 90]. In referring to classic tales of hubris such as the Greek myth of Prometheus and Francis Bacon's vision of "New Atlantis," von Wright attempted to mobilize cultural history and what he called the humanist attitude to life in order to evaluate the ways in which science and technology are used in society.

In this book, we present the history of science and technology as a tension between hubris and a hybrid imagination. We will focus on the social and cultural movements that have periodically emerged since the dawn of "Western" civilization to tame the hubris that has been generated as a central part of socio-economic development. From the Renaissance to the Enlightenment and from the romantic movements of the 18th and 19th centuries to the modernist and environmental movements of the 20th, social and cultural movements have served to counter the hubris that is so essential for scientific and technological development and thereby help guide it into more humane directions. By bringing people with different backgrounds and experiences together in pursuit of a common cause, these movements have provided sites, or contexts, for fostering a hybrid imagination, which, as we shall see, has proved crucially important for the further development of science, engineering and the broader society.

Hubris in History

o *The myths of Icarus and Prometheus*

o *The scientific revolution: "New Atlantis"*

o *Industrialization: "Prometheus Unbound"*

o *Atomic energy: "Science - The Endless Frontier"*

o *The arms race and the Apollo Mission*

o *Information society and green business*

A hybrid imagination can be defined as the combination of a scientific-technical problem-solving competence with an understanding of the problems that need to be solved. It is a mixing of scientific knowledge and technical skills with what might be termed cultural empathy, that is, an interest in reflecting on the cultural implications of science and technology in general and one's own contribution as a scientist or engineer, in particular. It can be thought of as an attitude of humility or modesty, as opposed to arrogance and hubris, in regard to scientific and technological development and, for that matter, to any kind of human activity. A hybrid imagination involves recognizing the limits to what we as a species and as individuals can do, both the physical limits and constraints imposed by "reality" as well as those stemming from our own individual limits of capabilities and knowledge. As such, a hybrid imagination is often manifested collectively, involving collaboration between two or more people even when it is not explicitly a part of a social or cultural movement.

The concept of a hybrid imagination is similar to what the planning theorist Bent Flyvbjerg, following Aristotle, has termed "phronesis," or judgmental knowledge. Aristotle famously identified the different types of knowledge in his *Nicomachean Ethics* as *episteme*, or theoretical-philosophical knowledge, *techne*, or practical-technical knowledge, and *phronesis*, or judgmental or ethical knowledge. Where the philosophers, like his own teacher Plato, tended to see the world in abstract terms, the technicians, who Aristotle thought of primarily as artists, focused on the concrete, seeing the

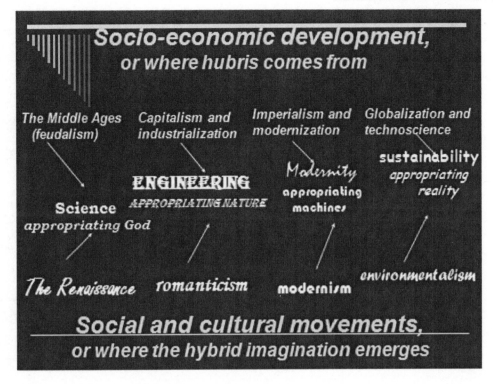

world in material terms. Phronesis, or "prudence," as the term is sometimes translated into English, was essential to both, according to Aristotle, but all too often neglected, and his work has long been used as a starting point for discussing the differences, as well as the interrelations between the various forms of knowledge [Flyvbjerg, B., 2001, McKeon, R., 1947].

In making his influential call for a phronetic social science that "matters," Flyvbjerg urged planning researchers and social scientists, more generally, to take the values of those they study into account, and to carry out research that focuses on value judgments and the ways in which values affect social action, not least in the planning of large infrastructural projects, or megaprojects like bridges and subway systems. Fostering a hybrid imagination in relation to science and technology has a similar ambition in regard to education. It emphasizes the importance for scientists and engineers to learn how to exercise judgment in assessing the cultural implications of their activity. It is the judgmental capacity to act wisely in context. By using the term, we want to suggest that there are different ways to respond to the challenges facing science and engineering in the contemporary world.

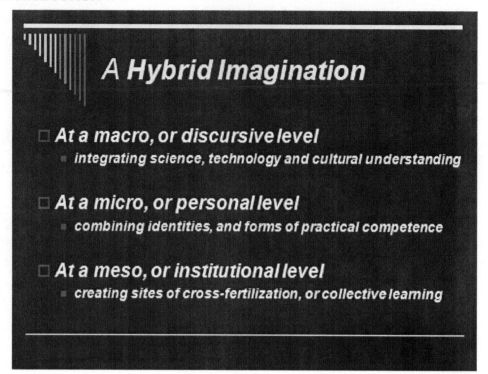

1.3   INTRODUCING PROCEED

There are many ways to characterize those challenges, and in the Program of Research on Opportunities and Challenges in Engineering Education in Denmark (PROCEED) that is being carried out between 2010 and 2013, we have taken our point of departure in three very different sorts of challenges. On the one hand, on a macro, or global level, there is what is often referred to as the sustainability challenge, or the overarching need for scientists and engineers – as well as for humanity in general – to relate to the problems brought to light in the debates about environmental protection, resource exploitation, and climate change. On the other hand, on a micro, or practitioner level, there are the challenges stemming from what we will be characterizing in this book as technoscience, the mixing in many fields of contemporary science and engineering of what has historically been termed science and what has historically been termed technology. And in between, at a meso, or institutional level, there are the various societal challenges, and a need for socio-technical competencies and a sense of social responsibility on the part of scientists and engineers, due to the permeation of science and technology into ever more areas of our economies, our societies, and our everyday lives.

Challenges Facing Science and Engineering

The sustainability challenge—how to deal with environmental problems, resource exploitation and climate change.

The societal challenge—how to deal with the permeation of our societies by science and technology in socially responsible ways.

The technoscientific challenge—how to combine scientific knowledge and technical skills with cultural understanding in new forms of competence.

The main response to these challenges can be characterized as hubristic, an external, or market-driven transdisciplinarity, by which the formation of a hybrid identity is imposed on scientists and engineers – and often self-imposed – in accordance with the interests of their sponsors or funders. The mixing is done for external reasons, and, as such, there is a tendency to become hubristic, to overemphasize the interests of the external sponsors and neglect the established quality standards and criteria of scientific and engineering work. This dominant response has been countered by what might be termed a residual or habitual strategy of "anti-hybridity" as many scientists and engineers attempt to resist change and reaffirm traditional academic and professional values and identities.

Response Strategies

The dominant , or hubristic strategy: "mode 2"
commercialization, entrepreneurship, transdisciplinarity.

The residual, or habitual strategy: "mode 1"
professionalization, expertise, (sub)disciplinarity.

An emerging, or hybrid strategy: "mode 3"
contextualization, engagement, cross-disciplinarity.

To foster a hybrid imagination is thus a way to respond to the challenges facing scientists and engineers in a spirit of critical engagement. It is neither to reject nor accept the new contextual conditions of science and engineering but is rather to creatively interact with them, questioning and assessing and reflecting on those conditions while learning to live with them.

1.4   THE TENDENCY TO HUBRIS

Already in the 1980s, Aant Elzinga noted how established epistemic criteria, that is, the ways in which truth claims are justified by scientists and engineers, were in a state of flux, as scientists and engineers increasingly found themselves in a condition of what he termed "epistemic drift:"

> …the process whereby, under strong relevance pressure, researchers become more concerned with external legitimation *vis-à-vis* policy bureaucracies and funding agencies than with internal legitimation via the process of peer review. This may be seen as a process of erosion of the traditional system of reputational control [Elzinga, A., 1985, p. 207].

Since then the traditional norms or values of scientists and engineers have been increasingly challenged by the transition to new ways, or modes, of doing research. To borrow a term from the French sociologist Bourdieu, P. [2004, p. 65], the "habitus" of science and engineering, a way of life based in distinct academic disciplines and professional identities, which provided a "collective capital of specialized methods and concepts," has been invaded by other forms of organization and ways of working.

In recent years, we have come to think of this process as a kind of hubris, in that so many decision-makers in universities, government, and business have not sufficiently taken into account the consequences of their decisions. In their efforts to make science and technology more effective providers of profitable innovations for the commercial marketplace, they have all too often neglected to consider the implications of what they were doing. Somewhere along the line, things have gone too far. In the words of Derek Bok, the former president of Harvard:

> Commercialization threatens to change the character of the university in ways that limit its freedom, sap its effectiveness, and lower its standing in society. … The problems come so gradually and silently that their link to commercialization may not even be perceived. Like individuals who experiment with drugs, therefore, campus officials may believe that they can proceed without serious risk [Bok, D., 2003].

There is, of course, nothing wrong with doing business. It is in the over-emphasis – and the relative marginalization of other reasons or motives for science and engineering – that the hubris comes in. If science and technology are to be nothing more than a part of commercial product development, a crucial element of "competition" in the global marketplace, then we are in serious trouble. Commercialization is based on the maximization of individual self-interest. The whole idea runs counter to doing things collectively for a common good. In relation to science and technology, it means that knowledge becomes a commodity, something to be owned or possessed for individual benefit. The pursuit of intellectual property rights makes sharing knowledge – and collective learning – difficult, if not impossible. And that is why it is so important that there be other ways for science, technology and the broader society to interact that emphasize justice, fairness, and cooperation, as has been discussed in other volumes in this book series.

A Tendency to Hubris

● ● ●

○ *transgressing established forms of quality control*
   ● *"a drift of epistemic criteria" (Elzinga)*

○ *transcending human limitations*
   ● *"converging technologies" (bio, info, cogno, nano)*

○ *neglecting the broader public, or social interest*
   ● *"academic capitalism": science and engineering in the private interest*

○ *(over)emphasis on commercialization*
   ● *propagation of competitiveness rather than cooperation*

The tendency to hubris is, in many ways, an essential part of science and technology – indeed, without an ambition to "play God" there can be no creativity – but as the Greek myths about Prometheus and Icarus pointed out so many centuries ago, there is a need to keep it under control if there is not be a tragic outcome. In the introduction to *Hubris and Hybrids*, hubris was characterized as the "if only" syndrome, "the eternal technical fixation that is deeply embedded in our underlying conceptions of reality:"

> If only we could develop an even better instrument of production and destruction, if only we could tame another force of nature to provide us with unlimited energy, then our wealth and our capacities – the values by which we measure progress – would be so much greater. More than two millennia after the sun melted the wings of Icarus for coming too close, we are still under the spell of hubris, trying to fly higher and higher [Hård, M. and Jamison, A., 2005, p. 5].

In most countries in the world today, it seems fair to say that attempts to foster science and engineering for social justice are quite marginal – if they exist at all – in comparison to the dominant ways of thinking about and practicing science and engineering. There is no real public space for serious discussion of the cultural implications of science and technology in most universities or, for that matter, in the media or anywhere else in the public sphere. Particularly in those countries, such as Denmark and the United States, where "technology assessment" became a part of the science and technology policy landscape in the 1980s and 1990s, there has been a very different political climate during the past decade. The places where technology assessment used to take place in the US, Denmark and elsewhere have been significantly reduced in size and influence, beginning with the closing of the US Office of Technology Assessment in 1994. While technology assessment had serious weaknesses in terms of the influence its practitioners actually had on science, technology and policy making, it did provide a space for reflection and public scrutiny. In such controversial areas as biotechnology, it helped provide accountability, transparency and a modicum of public participation in decision-making.

In any case, assessment has generally been replaced by various types of forecasting, or what have come to be called "foresight" activities, which provide analyses of potential markets and identify application areas that are most suitable for commercialization. Decades of telling stories of "entrepreneurship" and linking universities to industry have transformed the identities of many scientists and engineers. As such, there is a need to tell other stories about science, technology and society in order to balance the overall picture.

1.5 THE FORCES OF HABITUS

Not all scientists and engineers have accepted the new world of academic capitalism. A good many of them have reacted quite critically to the changing contextual conditions, and have sought to reaffirm a more academic, or professional, approach to science and engineering as a way to respond to the challenges. And while it certainly is valuable to uphold the importance of academic quality and professional standards, such responses tend to become anachronistic, in that they all too often merely reassert the traditional norms of academic life and professional behavior, without recognizing that those norms and values have, to a large extent, become outmoded [Christensen, S.H. and Ernø-Kjølhede, E., 2006]. By force of habit(us), the responses all too often become a defense of a tradition or a customary way of life that is simply no longer viable rather than a meaningful strategy of response.

As part of this strategy, it has become popular to refer to the norms of science, which were influentially formulated in the 1940s by the American sociologist Merton, R. [1942]. These have long been seen by many natural and social scientists, as well as engineers and large segments of the general public, as well, as core values in science and engineering. The norms of communalism, universalism, disinterestedness and, not least, organized skepticism continue to be seen as defining features of science, even though the practice of science has fundamentally changed since Merton characterized them. As we shall see in chapter six, these norms have been challenged during the second half of the 20th century: first by the coming of "big science" and, with it, what might be called bureaucratic norms in the decades following the second world war and, more recently, by the coming of technoscience and the explicitly commercial norms that have come along with it. But the Mertonian norms continue to be propagated and considered part of the identities of scientists and engineers, particularly in relation to contentious issues such as climate change.

The Forces of Habit(us)

● ● ●

- o *Technoscience primarily seen as providing new opportunities for scientists and engineers*

- o *Taught by restructuring established scientific and engineering fields: multi- or "subdisciplinarity"*

- o *Politics and the rest of society left largely outside of research and education: "outsourcing" of ethics*

- o *A continuing belief in separating experts and their knowledge from contexts of use*

In order to meet the challenges facing science and engineering in the world today it is not sufficient to reaffirm a traditional faith in reason and truth and reassert the importance of a largely outmoded form of imagined academic community. There is instead a need to foster a hybrid imagination, connecting science, technology and society in new ways, by combining scientific knowledge and technical skills with cultural understanding, or empathy.

1.6   THE STRUCTURE OF THE BOOK

It is the underlying argument of the chapters that follow that if science and technology are to help solve problems rather than cause new ones, then they need to interact with the rest of society in more appropriate ways than is currently the case.

After outlining the argument and defining concepts in the next chapter, the third chapter traces the emergence of "modern" science and technology from the hybrid imaginations of the artist-engineers of the Renaissance to the scientific revolution of the 17th century and onto the "age of enlightenment" of the 18th century. The fourth chapter focuses on the 19th century and the importance of the romantic and cooperative movements in the first half of the century and the socialist and populist movements in the second half in fostering hybrid imaginations in the making of science and technology. In the course of industrialization, science and technology were remixed in new combinations, giving rise to new branches of industry and new discourses, institutions and practices.

The fifth chapter continues the story into the third of what the economic historian Joseph Schumpeter called "long waves" of industrial development, by discussing the making of modernity and the role of modernist and anti-colonial movements in the broader processes of modernization in the first half of the 20th century. Chapter six brings the story up to date by discussing the emergence of new societal and technological challenges relating to science and engineering during the second half of the 20th century and the role played by the social and cultural movements of the 1960s and 1970s. Finally, the concluding chapter takes a look at the ongoing and highly contentious process of "greening" science and technology that has served as my main research focus through the years, and where the contending forms of hybridization – "top-down" green business versus "bottom-up" environmental activism – have both been resisted by the anti-hybridity of the climate skeptics and anti-environmental traditionalists both inside and outside of the universities. As such, the book attempts to place science and technology in a broad historical and cultural perspective.

CHAPTER 2

Perceptions of Science and Technology

> There is a sort of poverty of the spirit which stands in glaring contrast to our scientific and technological abundance. The richer we have become materially, the poorer we have become morally and spiritually. We have learned to fly the air like birds and swim the sea like fish, but we have not learned the simple art of living together as brothers.
>
> Every man lives in two realms, the internal and the external. The internal is that realm of spiritual ends expressed in art, literature, morals, and religion. The external is that complex of devices, techniques, mechanisms, and instrumentalities by means of which we live. Our problem today is that we have allowed the internal to become lost in the external. We have allowed the means by which we live to outdistance the ends for which we live.
>
> Martin Luther King Jr., "The Quest for Peace and Justice" (1964)

2.1 THE IMPORTANCE OF CONTEXT

It is nearly half a century since Martin Luther King Jr., on receiving the Nobel Peace Prize for his leadership of the American civil rights movement, pointed to the division in the modern world between means and ends, external things and internal values. Science and technology had brought progress and prosperity, but social injustice remained. In the remaining years of his life, before he was killed in 1968, King would be one of the millions around the world to protest against the misuse of science and technology in the Vietnam war. As we shall see in chapter six, the protests against the militarization of science and technology would be one of the sources of inspiration for the emergence of the social and cultural movements in the 1960s and 1970s, which would foster alternative approaches to science and technology as part of their activity.

During the intervening decades, following the demise of those movements in the 1980s, commercial networks, or so-called systems of innovation linking academic engineers and scientists with business firms have become the main institutional frameworks for scientific and technological development. This means that scientific and technological development has come to be governed and managed, to a much greater extent than in the past, by its sponsors, or funders. The changes that have taken place in the contexts of knowledge production have served to turn the making of much, if not most, science and technology into competitive rather than cooperative activities, and as a result, it has become ever more difficult to bridge the gap between scientific and technological development and what Martin Luther King Jr. termed the quest for peace and justice.

As opposed to more traditional, or disciplinary forms of knowledge-making that were prevalent in the past, scientists and engineers in the contemporary world must take more directly into account the interests of those who provide the funding and support their work. And that is why it is so important for scientists and engineers in their education to learn something about the contexts in which they will be working.

2.2   THE STORY-LINES OF SCIENCE AND TECHNOLOGY

It can be useful before going any further to attempt to distinguish, in an ideal-typical fashion, some of the more significant approaches to understanding the contexts of science and technology. There is a wide range of meanings that science and technology have in society, and the attempts to understand them make use of different terminologies and intellectual traditions that correspond to the different functions or roles that science and technology play in society: economic, social, and cultural [Jamison, A. and Hård, M., 2003].

The central meaning of science and technology, at least as they have developed since the mid-19th century, has been commercial, that is, making money and doing business. Science and technology have come to mean the transformation of "inventions" or new scientific discoveries into "innovations" that can profitably be sold on the commercial marketplace. When the contexts of science and technology are discussed, whether by academics, politicians, journalists, or just plain ordinary "consumers," attention tends to be directed primarily to the activities of companies and corporations and to the scientists and engineers who have achieved commercial success, for these are generally considered to be the main sites, or contexts in which innovation takes place. The relevant contextual knowledge is thus almost exclusively economic and managerial, with a focus on identifying and analyzing the "entrepreneurial" skills that are considered crucial for bringing scientific discoveries or technical inventions to market.

While the importance of commercial science and technology can certainly not be denied, there are nonetheless other meanings of science and technology that are at least of equal importance, if not moreso. Indeed, it can be suggested that the dominance of the commercial, or economic meaning has led to a relative neglect of what might be termed the social and cultural meanings of science and technology. Many forms of scientific and technological development are intrinsically social, in the sense that they are attempts to apply technical skills and scientific ingenuity to the solving of social problems and/or the resolution of social conflicts. The aim has been to provide a kind of structural, or what is often referred to as "infrastructural," contribution to a unit of social organization, be it a

What do science and technology mean?

● ● ●

o *making money, doing business:*
ideas/inventions > innovations > products > profits

o *solving problems, controlling society:*
interests > actors/networks > artifacts > systems

o *empowering people, using things:*
movements > institutions > values/behavior

city or a nation-state or a society. It has usually involved one of another form of system-building or systemic thinking, by which various component parts – both scientific, technical and non-technical, both non-human and human – are brought together into a larger coordinated effort.

The interesting questions in relation to contextual knowledge in this kind of science and technology revolve around the social sector or domain in which these processes take place and the particular kinds of competence or expertise that are required. In these areas of science and technology, the task has generally been to transform an idea, plan, design, or vision into material manifestations and, by so doing, help to "fix" a problem or resolve a conflict that has been identified as socially significant – public health, energy supply, electricity distribution, environmental protection, transportation planning, etc. This can be thought of as the social, or professional meaning of science and technology.

Even less recognized than the social in relation to the dominant economic meaning is a third ideal-typical meaning, which can be characterized as cultural and which represents the ways in which people have cooperated with one another to learn how to deal with the basic problems of their lives. This meaning has been given far less attention than the other two, due perhaps to its intrinsic diversity and variety, as well as to what might be called its "situatedness" or particularity; it is hard to aggregate or theorize about these forms of science and technology, but they are nonetheless of central

...and what stories can we tell about them?

o *stories of economic innovation:*
   focus on entrepreneurs, firms and competitiveness

o *stories of social construction:*
   focus on actors, networks and consensus-building

o *stories of cultural appropriation:*
   focus on change agents, movements and contention

importance for many areas of human existence. The interesting questions in relation to contextual knowledge revolve around the processes of what might be called collective knowledge-making, or cultural learning, and, more specifically, the capacity to use science and technology in appropriate ways. This can be considered the cultural, or human meaning, of science and technology.

2.3   TEACHING CONTEXTUAL KNOWLEDGE AT AALBORG

Distinguishing contextual knowledge in this way draws on many years of teaching non-technical material to science and engineering students, most recently, at Aalborg University in Denmark. Like many other universities that were created in the 1970s, under the influence of the social and cultural movements of the times, Aalborg University attempted to develop a more "relevant" form of education than was then being offered by the established universities.

From the outset, Aalborg University has based all of its undergraduate teaching programs on a combination of problem and project-based learning, with formalized courses playing only a subsidiary or supportive role. For the most part, the students are taught their subjects by carrying out semester-long projects in groups, and the task of the teacher is primarily to advise the students, rather than instruct them.

In the science and engineering fields, project work in the first year has included, since the early 1980s, a certain amount of what has come to be referred to as contextual knowledge. The particular way in which this knowledge is taught and included in the student projects varies from field to field, and it has also varied from year to year, depending on who is doing the teaching and, not least, on the relations between the main, scientific/technical advisers, who are responsible for the project work as a whole and the contextual advisers, who, for the most part, come from outside the particular field of study. Most of the contextual advisers have a social scientific and/or humanities education, and there has thus been a wide range of approaches to contextual knowledge that have been presented in the supportive courses that are given, and then put to use in the student projects.

Approaches to Teaching Contextual Knowledge in Aalborg		
Strategy	Market orientation, Transdisciplinarity	Professionalization, (sub)disciplinarity	Cultural awareness, cross-disciplinarity
Rationale	contextual knowledge is for cultivating entrepreneurship	contextual knowledge is for habituating students in a field	contextual knowledge is for fostering a hybrid imagination
Storyline	Economic innovation, technological development	Social construction, science and technology in action	Cultural appropriation, narratives of use
Main contents	Innovation and management studies, market analysis	Philosophy of science and S&T studies, actor and network analysis	History of S&T and cultural studies, technology assessment

The most common approach to contextual knowledge has been to provide a kind of supplementary, or add-on knowledge, usually aimed at offering the students knowledge of some of the "market" conditions that affect their particular engineering or scientific field. Typically, the lectures and advising focuses on managerial issues and "entrepreneurship," and the project work often involves one or another form of market analysis of the particular technical or scientific product that the students are learning how to design and/or build in their projects. This tends to be an instrumental approach to contextual knowledge, rather than a reflective or theoretical approach.

A second approach that is used in Aalborg provides more of a complementary or extra-curricular knowledge, offering students an opportunity to reflect on the underlying values and paradigmatic assumptions of their scientific or technological field as a way of preparing for their future professional roles. The courses usually offer an introduction to the philosophy and/or sociology of science and technology, presenting some of the different schools, or positions, as well as some of the methods of analysis that have been developed in science and technology studies. The social construction of technology, or SCOT, perspective, as developed by Wiebe Bijker, Thomas Hughes, Trevor Pinch and others [Bijker et al., 1987] has been especially popular in recent years.

A third approach, and one that has been developed in recent years in the educational programs in medialogy and nanotechnology, is to connect, as much as possible, the technical-scientific components of the project work to broader cultural awareness, and to mix something of the more instrumental ambition of the market-oriented approach with the reflective ambition of the professional approach. The lectures introduce the students to the cultural history of science and technology, as well as to some of the public debates that have taken place in relation to science and technology. Students have also been introduced to political and ethical aspects of contemporary science and technology and to some of the methods that are used to study public attitudes and opinions about scientific, technological, and environmental issues. The contextual advising of the project work has been seen as a way to help the students learn how they might address and, at best, assess the political, cultural and/or environmental implications of their particular scientific-technical project. In one case, students integrated a comprehensive examination of the climate change debate into their scientific and technical project in nanotechnology on "raspberry solar cells" and in others, students conducted ambitious surveys of public attitudes to nanotechnology as a part of their project work [Jamison, A. and Mejlgaard, N., 2010]. In still others, students making computer games or interactive websites carried out focus group interviews with the potential users of their technical projects and also tried to bring a cultural or political motivation into their project work. One made an interactive website about Auschwitz as a way to counter the lack of public awareness of the Holocaust, and another made a computer game about gender roles.

In the following sections, the story-lines on which these different approaches to contextual knowledge are based will be briefly presented.

2.4   THE ECONOMIC, OR INNOVATION STORY-LINE

Karl Marx was perhaps the most influential of a new breed of economic and social scientists in the mid-19th century who focused attention on the role of science and technology in economic life. Marx saw in the coming of modern industry and in the use of science and technology in the economy an epochal shift in human history. Marx's insights into the economic significance of science and technology have been highly influential in shaping the widespread belief in science-based progress. They became an important part of the political consciousness of those who created social-democratic and communist parties, and they also formed a central part of economic history, as it developed in the late 19th and early 20th centuries as a kind of hybrid academic field combining economics and history.

One of those who helped turn the Marxian insights into a story-line of economic innovation was Joseph Schumpeter, who coined the term "creative destruction" that has since provided an underlying narrative trope, or metaphor, for science-based industrial development. "The essential point to grasp is that in dealing with capitalism we are dealing with an evolutionary process," he wrote toward the end of his life in *Capitalism, Socialism and Democracy*. And at the core of the evolutionary process that was capitalism was the process of innovation that "incessantly revolutionizes the economic structure from within, incessantly destroying the old one, incessantly creating a new one" [Schumpeter, J., 1975, p. 82, 83].

Drawing on the work of a Russian economist, Nikolai Kondratiev, Schumpeter developed a model of business cycles, or "long waves," in which the process of innovation played a central role [Freeman, C. and Louçá, F., 2003]. Schumpeter's ideas have been formative for the ways in which economists and economic historians and, not least, those in the sub-fields of technology management and innovation studies have since come to characterize and analyze the contexts of science and technology. At the beginning of each wave, a cluster of radical innovations are seen to propel a new upswing in industrial expansion as they are spread, or diffused, in the economy. New companies and branches of industry that are based on the radical innovations emerge to replace the companies that had grown up in the previous waves.

Christopher Freeman's book, *The Economics of Industrial Innovation* [Freeman, C., 1974], and later his analysis of postwar Japanese technological development, *Technology Policy and Economic Performance: Lessons from Japan* [Freeman, C., 1987], were central texts in the revival of innovation economics in the 1970s and 1980s. Freeman discussed the different innovation strategies that companies pursue and pointed to the importance of strong ties, or linkages between companies, government agencies, and universities in what he termed the Japanese system of innovation. The idea of a national system of innovation was applied in Denmark, as well, by a group of economists at Aalborg University, who told the story of Danish industrialization in the 19th and 20th centuries as a process of creating an "agricultural-industrial complex" or development block, drawing on particular kinds of engineering activities [Lundvall, B., 1992].

Since then, economists and historians have discussed systems of innovation, both in particular countries, economic branches and fields of science and technology. There are now departments of innovation studies at many business schools, and the story-line of innovation has come to provide the dominant way in which technology is discussed, both in the academic and more policy-oriented literature. There is also, of course, a more popular writing about technology in the large number of works on particular "success stories" – of products, companies, and individual inventions, from the personal computer to the Internet, from Microsoft to Google.

The ways in which these stories are told follows a typical pattern, which can be characterized as a form of technological determinism, according to which new, radical innovations – in our day, primarily in information technologies, genetic engineering, and nanotechnology – are claimed to be the central factors behind economic growth and "competitiveness."

2.5   THE SOCIAL, OR CONSTRUCTION STORY-LINE

While economists and economic historians, and the stories of innovation that they like to tell, tend to dominate both the public understanding, as well as the academic study, of science, technology and society, a second significant story-line or narrative approach has emerged within sociology and

● ● ●   **The Story-line of Economic Innovation**

also known as "technological determinism":
science and technology cause changes in society

o *Steam engines and textile machines give us industrial society*

o *Automobiles and telephones give us modern society*

o *Television and computers give us information society*

o *Atomic energy and genetic engineering give us risk society*

o *Info, bio- and nanotechnologies give us knowledge society*

philosophy and the sub-field of science and technology studies. The roots of this work can be traced back to some of the early historians of engineering in the 19th century, such as Samuel Smiles, who wrote biographies of the bridge-builders and railway engineers who constructed the industrial society, as well as to the early social theorists and sociologists, such as Max Weber, Emile Durkheim, and Robert Merton, who had begun in the late 19th and early 20th centuries to consider some of the social factors that were involved in the development of science and technology.

Weber emphasized the processes of rationalization and bureaucratization that were at work in the formation of modern societies, and which had a major influence on science and technology, especially perhaps in what are now called infrastructural projects. He also wrote about the underlying values, or norms, of behavior in many areas of social life, linking social activity to what he termed an underlying ethical system, or ethos. His famous book, *The Protestant Ethic and the Spirit of Capitalism* [Weber, M., 2001] stressed the religious, or moral basis of capitalism in the interest in technical improvement that was so much a part of the new forms of Christianity that emerged during the Protestant Reformation of the 16th century.

The story-line of construction includes both a macro, or discursive, level, at which overarching principles of social structure and organization are discussed (from the "iron cage" or rationalization process of Max Weber to the technological rationality of Herbert Marcuse and the power discourses of Michel Foucault), a micro level, at which particular projects are carried out, and an intermediary, or meso level of what are often called "large technical systems." The relevant contexts depend on the level of story-telling, but they tend to be abstract social structures at the macro level, individual actors and networks of individuals at the micro level, and institutions and social organizations at the meso level.

Since the 1980s, the French philosopher and anthropologist Bruno Latour and the Dutch engineer-turned-sociologist Wiebe Bijker have been among the most active in developing the story-line of construction. Latour has emphasized the ways in which scientists and engineers have constructed "actor-networks" that bring together human and "non-human" elements in their various projects. It is, as he has characteristically put it in the title of one his books, "aramis, or love of technology" that forms a kind of core meaning of engineering work, and the kind of contextual knowledge that he has been so influential in developing has focused on the ways in which this love has been put into practice, not always with positive results [Latour, B., 1986, 1996, 2005].

● ● ● | **The Story-Line of Social Construction**

Actors "co-construct" scientific facts and technological artifacts with non-humans to satisfy social interests

o *An interest in mobility and individual freedom leads to the bicycle and the automobile*

o *An interest in protecting non-human "nature" leads to technologies of environmental control*

o *An interest in a technologically-mediated, or virtual reality leads to new information technologies*

Wiebe Bijker and the American Thomas Hughes, on the other hand, have provided a number of case studies of key "system-builders" or network-makers, seeking to uncover the ways in which scientists, engineers and other actors through their professional activities actually go about shaping social institutions and organizations. Hughes has contrasted the "networks of power" that were involved in the development of electricity systems in Europe and the United States [Hughes, T., 1983], and Bijker has elucidated the social interests and technological frames that were at work in a number of different fields [Bijker, W., 1995].

Many other analyses of actor-networks and social construction have been developed during the past twenty years, emphasizing the "co-construction" of science, technology and society by focusing attention on how scientists and engineers turn social interests into "facts" and "artifacts." The basic idea is to study scientists and engineers in action, by using ethnographic methods and applying a constructivist epistemology.

The story-line of construction emphasizes social processes rather than economic ones, and its story-tellers employ a language or vocabulary of sociology, anthropology and social philosophy to recount their tales of networking, negotiation and mediation. The stories that are told in this form of contextual knowledge are often those that take place at the interface or meeting place between the worlds of business, government and academic life, or what are increasingly referred to as the contexts of "governance." The emphasis is on the "actor-networks" involved in the construction of a scientifically and technologically mediated reality. The expertise or professional competence of scientists and engineers is thus not seen as purely technical or scientific; there is also a kind of social competence, or social capital that is necessary.

2.6   THE CULTURAL, OR APPROPRIATION STORY-LINE

While the economic, or innovation, story-line is by far the most dominant, the social, or construction story-line has become ever more influential in recent years, especially in the arenas of policy-making and government. Both focus on the production of science and technology, and have thus tended to pay limited attention to the more cultural aspects of technology and science, which will be the main focus of attention in the chapters that follow. A main source of inspiration for this story-line was the American writer, Lewis Mumford, and, in particular, his classic work, *Technics and Civilization*, from 1934, which was one of the first to discuss the cultural preconditions, as well as the cultural consequences and contradictions of science and technology. Later in his life, he became one of the main critics of the so-called military-industrial complex in the United States, which he saw as a new kind of authoritarian technology, what he termed the megamachine [Mumford, L., 1970].

More recently, the British cultural historian and theorist Raymond Williams has written about the relations between technology and broader cultural processes in a number of books that have contributed to the creation of the academic field of cultural studies. In his writings, Williams emphasized how the idea of culture had emerged in the 19th century as a "record of our reactions, in thought and feeling, to the changed conditions of our common life… Its basic element is its effort at total qualitative assessment"[Williams, R., 1958, p. 285]. For Williams, the idea of culture has served as a critical counterpoint to technological and scientific development:

> The development of the idea of culture has, throughout, been a criticism of what has been called the bourgeois idea of society. The contributors to its meaning have started from widely different positions and have reached widely various attachments and loyalties.

● ● ● **The Story-line of Cultural Appropriation**

The meanings of technologies come with use, and become matters of contention and debate

○ *Using machines in factories transforms the meaning of production, but also destroys pre-industrial forms*

○ *Using automobiles transforms the meaning of society, but also destroys the environment*

○ *Using info-, bio- and nanotechnologies transforms the meaning of humanity, but also challenges many traditional values*

But they have been alike in this, that they have been unable to think of society as a merely neutral area, or as an abstract regulating mechanism. The stress has fallen on the positive function of society, on the fact that the values of individual men are rooted in society, and on the need to think and feel in these common terms.

Another influential writer has been the literary historian, Leo Marx, who was a pioneer in investigating the artistic and literary representations of science, technology and engineering in his important study, *The Machine in the Garden*, from 1964. Marx's student, David Nye, has been one of the most prolific contributors to the story-line of appropriation, in a series of books on the ways in which electricity and other forms of power have been used in different ways by different people. His recent book, *Technology Matters*, provides a highly readable introduction to this way of discussing technology and its cultural contexts [Nye, D., 2006].

This third story-line focuses on very different contexts or social locations than the other two, often telling stories about broader social and cultural movements, as we will be doing in the chapters that follow. Focusing on these contexts brings out the ways in which science and technology have often first developed in informal and temporary public spaces carved out by social and cultural movements, before becoming institutionalized in the more formal or established contexts that are of interest in the other story-lines.

The story-line of cultural appropriation, as we will be presenting it here, builds on the cognitive approach to social movements that was developed in the 1990s together with the sociologist Ron Eyerman. An investigation of the relations between environmental movements and knowledge [Jamison et al., 1990] led to the recognition that social movements are more than "merely" political phenomena, and that in order to understand their contribution to scientific and technological development, it was necessary to conceptualize them in a new way.

The cognitive approach makes use of the terms "cognitive praxis" and "movement intellectuals" to emphasize the role of knowledge-making in social movements and to characterize the people who are most actively involved [Eyerman, R. and Jamison, A., 1991]. Cognitive praxis is defined as the linking, or integration, of ideas, ideologies, and/or world view assumptions (a cosmological dimension) to particular activities or forms of action, including technical development, information dissemination and practical demonstration of both protest and constructive alternative (a technological dimension). The movement is seen as providing an organizational dimension, a public space, for integrating the cosmology and the technology in processes of collective learning, and it is their cognitive praxis that makes social movements particularly important in the constitution and reconstitution of science and technology [Jamison, A., 2006].

In a book on music and social movements [Eyerman, R. and Jamison, A., 1998] the cognitive approach was broadened to encompass cultural practices and, by so doing, focus attention on the role that the "mobilization of tradition" plays in the collective activities of many social movements. It is the mobilization and (re)invention of different traditions of ideas, beliefs and ideologies that often plays an important role in attracting active participation and involvement in social movements. In the cognitive approach, social movements are thus seen to provide spaces in the broader culture for new forms of science and technology to emerge.

Historically, these forms of science and technology have been part of broader social and political struggles, from the Reformation of the 16th and 17th centuries through the social movements of the 19th and 20th centuries and into the present. As we shall see in chapter four, one of the first environmental scientists, Henry David Thoreau, took active part in the movement to abolish slavery in 19th century America, and one of the founders of interior design, William Morris, was an active member of the socialist movement, as well as a professional artist and designer. In the early 20th century, as we shall see in chapter five, the modernist movements in Europe and the United States provided sites for the emergence of new combinations of science and engineering, and art and technology as well as influential approaches to urban planning, architecture and design. Similarly, as we shall see in chapter seven, in the environmental movements of the 1970s, grass-roots forms of engineering provided "utopian" or radical examples of appropriate technology that have since developed into significant branches of industry [Dickson, D., 1974].

Particularly influential was how, within the context of the opposition to nuclear energy, many professional scientists and engineers joined forces with environmental activists to experiment with alternative forms of energy. In Denmark, as a part of the movement against nuclear energy, an organization for renewable energy was created that provided a space, or cultural context in which people could learn how to build wind energy power plants and solar panels. Like similar activities in other countries, these forms of science and engineering were a kind of democratic, or grass-roots, innovation process, and like other movements today, in organic agriculture, alternative health care, sustainable design and architecture, they open scientific and technological development to popular, or public, participation.

In what follows, we will present the history of science and technology in relation to what Raymond Williams in the 1970s termed contending "cultural formations." For Williams, cultural change involves the emergence of new "structures of feeling," new mixtures of ideas and practices, or what Williams termed "social experiences in solution"[Williams, R., 1977, p. 133]. Emerging cultural formations or structures of feeling are subjected to two sorts of pressures: incorporation from the dominant cultural formations, and reaction from residual cultural formations.

In this sense, a dominant cultural formation in regard to science and engineering seeks to incorporate all scientific and technical developments into commerce, and as we have discussed in the introduction, there is a tendency to hubris in these incorporation processes. The various forms of resistance or reaction – the defence of traditional values in both science and engineering – can be seen as residual cultural formations, trying to adapt scientific and technological developments to older, more habitual ways of life. Where the one attempts to turn all new ideas and inventions into business, the other seeks to uphold the norms and values that have traditionally been associated with science and engineering.

Contending Cultures of Science and Engineering			
	Residual	*Dominant*	*Emergent*
core values	academic professional	commercial entrepreneurial	cooperative flexible
discursive tradition	rational analytical	utilitarian pragamatic	systemic holistic
type of knowledge	abstract mathematical	empirical experimental	synthetic contextual
organizational form	discipline-based research groups	market-driven networks	change-oriented aliances
form of education	scientific philosophical	managerial technical	integrative cross-cultural
type of learning	"by the book"	"by doing"	"hybrid imagining"

It is somewhere in between, and in a kind of struggle with both the dominant and the residual cultures, that what Williams termed an emergent cultural formation can be identified in regard to science, technology and engineering. Both in terms of core values, organizational forms, types of knowledge, and forms of education new approaches to science and technology can be seen to emerge in periods when social and cultural movements bring into question the established forms, both dominant and residual, and foster what we will be calling in the chapters that follow, a hybrid imagination. As such, the history of science and technology, as we will be tracing it, is a recurrent cyclical process of movement making versus institution building, of change agents struggling against the forces of incorporation and hubris, on the one side, and those of resistance and habitus on the other. It is in the spaces, or culural contexts, in between hubris and habitus that a hybrid imagination, which has proved so essential to the development of science and technology, has been fostered.

CHAPTER 3

Where Did Science and Technology Come From?

If we are to call any age golden, it must be our age which has produced such a wealth of golden intellects. Evidence of this is provided by the inventions of this age. For this century, like a golden age, has restored to light the liberal arts that were almost extinct: grammar, poetry, oratory, painting, sculpture, architecture, music, the ancient singing of songs to the Orphic lyre, and all this in Florence. The two gifts venerated by the ancients but almost totally forgotten since have been reunited in our age: wisdom with eloquence and prudence with military art.

Marsilio Ficino (1492) (quoted in Brown, A. [1999, p. 1])

3.1   THE RISE OF THE WEST

Science and technology as we call them today came into the world as an integral part of the rise of Western civilization, which took place in the period stretching from the Renaissance of the 15th and 16th centuries to the Enlightenment of the 17th and 18th centuries. The subsequent 200 years, from the late 18th to the late 20th century, are generally considered to be a period in which the "West" dominated the rest of the world, not least because of the superiority of the science and technology that Europeans, and, eventually, the transplanted Europeans in North America had developed.

Before that time, science and technology did not exist, or, more precisely, they did not exist in the same form or have the same meanings as they have had during the past 250 years or so. There was instead a great deal of regional variation, and in most places, there was a rather large gap between what came to be called science and what came to be called technology.

In medieval Europe, as in most other parts of the world, the production of scientific "facts" had been separated, both discursively, institutionally and practically, from the production of technical "artifacts." There was both a physical distance and a cognitive distinction between the theoretical philosophical knowledge that the ancient Greek philosopher Aristotle in his *Nicomachean Ethics* had termed "episteme" and the practical-technical knowledge that he had termed "techne"[McKeon, R.,

A Brief History of Technology and Science

● ● ●

o *Ancient, or Traditional wisdom, up to about 1600*
  - **spiritual knowledge, distinctive regional modes**
  - **gap between theory (episteme) and practice (techne)**

o *Modern, or Western science, from about 1660 to 1980*
  - **instrumental, rational, universal knowledge**
  - **functional interdependence of science and technology**

o *Global, or Technoscience, from about 1980*
  - **multiple forms of knowledge, commercial networks of innovation**
  - **combinations of science and technology**

1947]. The one tended to be made in a world of "free men" while the other was often the work of slaves, or at least people lower down in the social hierarchy. Neither on an institutional nor intellectual level were their active connections between the "theorists" and the "technicians" in any part of the world, although there certainly were individual people who were interested in both kinds of knowledge.

As for "phronesis" – the moral or judgmental knowledge that Aristotle had considered most important of all – it had more or less been outsourced, in Europe as elsewhere to politicians, priests and judges. Throughout the world, scholars, craftsmen, and politicians – the makers of the different kinds of knowledge – tended to live and work in distinct lifeworlds, separated from each other both physically and mentally. There were deep-seated intellectual, or cultural traditions that served to guide scientific inquiry, technical development, and ethical reflection into very different directions, what in the West have been seen as mathematical-theoretical, empirical-practical, and systemic-synthetic. The makers of knowledge in the different life-worlds tended to have very different roles in society, and they tended to have very little interaction with one another. It was in medieval Europe, and in medieval Europe alone, that those gaps began to be bridged by people with what might be termed a collective, or shared hybrid imagination, bringing theory, practice, and politics together into modern, or Western science and technology.

Historians give different reasons for how and why this happened, depending on their approach and interests, or what we have characterized in chapter two as their "story-line." The dominant view, corresponding to what we have termed the economic, or innovation story-line, is that something called "capitalism" or a market economy is primarily responsible for the rise of the West, and, more specifically, that it was the growth of new kinds of economic activities, based on particular technical innovations and scientific discoveries that led to the European and, later, North American conquest of the rest of the world. In this story-line, scientific and technologically-driven economic developments are seen to have played a determinant role in the rise of the West and the subsequent age of imperialism in the 19th and 20th centuries (cf. Landes, D., 1969, Mokyr, J., 1990).

In the early middle ages, the development of new agricultural techniques, in particular a heavier plough and the use of the horse as a farming animal, led to an enormous increase in food production, which, in turn, made possible the growth of towns. In the 12th and 13th centuries, there was what has been termed an "industrial revolution of the middle ages" when a new range of manufacturing and construction techniques contributed to the rise of inter-urban trade and commerce (cf. Gimpel, J., 1976, White, L., 1962). There was also a development of military technology, especially artillery that gave rise to large standing armies as the nation-states of Europe freed themselves from the Roman Catholic Church. Later, the invention of the printing press and improvements in the techniques of mining and metallurgy – as well as in the science and technology of ship-building and navigation – made it possible for European explorers to journey around the globe and eventually exploit other people and their natural resources for the benefit of European industrialization (cf. Cipolla, C., 1965, Eisenstein, E., 1983). In the 19th century, the railroads and the scientific and technological achievements that came with them formed the material basis for imperialism (cf. Headrick, D., 1981, Landes, D., 1998).

Other historians, corresponding to what we have termed a story line of social construction, give primary responsibility to individual actors and their networks – the kings and popes and noble-men and the scientists and engineers who they sponsored. It was because of the way that the rulers ruled and the kinds of institutions that they established that Europe came to dominate the world. In particular, the laws and regulations and organizations that were put in place, not least in encouraging and sponsoring scientific and technological development led to a new kind of innovative and exper-imental spirit, or ethos (cf. Huff, T., 2003, Weber, M., 2001). In relation to science and technology, the emphasis has thus been on the social conditions and the social institutions that made it possible for certain great men of genius to develop new frameworks for constructing facts and artifacts – the mechanical philosophy and the experimental method – in the so-called scientific revolution of the 17th century and the subsequent professionalization of science and engineering in the 18 and 19th centuries (cf. Shapin, S., 1996, 2008).

As the dominance of the "West" has faded into history and the rise of Western civilization is no longer seen as an unabashed success story, other stories have begun to be told about the history of science and technology. According to what we have called a cultural, or appropriation story-line, science and technology are seen to have been a decidedly mixed blessing, with both positive and

● ● ● | ***The Economic Story-Line***

- ○ *The agricultural revolution, ca 600-900*

- ○ *The urban migration and growth of towns*

- ○ *The industrial revolution of the 12th century*

- ○ *Exploration and international trade*

- ○ *Mining and the rise of capitalism*

- ○ *Industrialization, railroads and imperialism*

negative consequences for humans, as well as for the non-human "environment." As such, it has been important to examine the historical record anew, from a fresh perspective, looking both for the good news and the bad news, the strengths as well as the weaknesses of modern science and technology.

What is of most interest has been not so much the triumphal rise of the West as the cultural contexts and movements in which the economic and social changes took place (cf. Jacob, M., 1988, Merchant, C., 1980, Webster, C., 1975). In the words of Frances Yates, whose works have been particularly influential in this regard:

> It may be illuminating to view the scientific revolution as in two phases, the first phase consisting of an animistic universe operated by magic, the second phase of a mathematical universe operated by mechanics. An enquiry into both phases, and their interactions may be a more fruitful line of historical approach to the problems raised by the science of today than the line which concentrates only on the seventeenth-century triumph [Yates, F., 2002, p. 452].

In what follows, we want to try to bring this cultural "background" into the foreground, for it is our contention that it was primarily through the ideas and practices that were formed in broader social and cultural movements that Europe emerged out of the "dark ages" and into the modern era. These movements represented a wide-ranging attempt to revive a once great European civilization that had gone through centuries of decline, and at their core were movement intellectuals with hybrid imaginations, mixing the ideas and ideals of Greek and Roman antiquity with the various technical practices and experiences that had developed in the "middle ages" and in other parts of the world. As these movements were transformed into the discourses, institutions and practices of modern, or Western science and technology, their "cognitive praxis" was fragmented into the contending cultures that we characterized in chapter two.

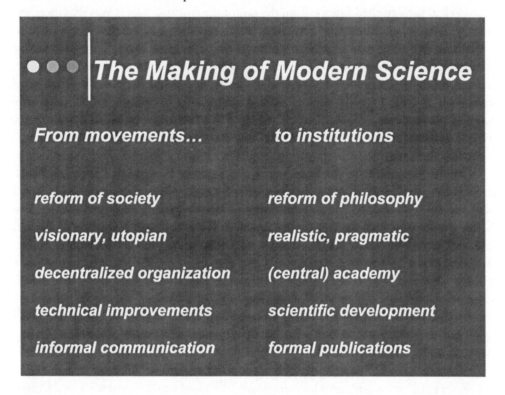

Seeing the development of science and technology in terms of a story-line of cultural appropriation focuses attention on the social and cultural movements of early modern Europe and on their knowledge-making activities. Our claim is that the discourses, institutions and practices of modern, Western science and technology grew out of seeds planted in social and cultural movements.

3.2   THE MEDIEVAL TRANSITION

In the centuries that followed the "fall" of Rome, there was an institutional separation in Europe as elsewhere between science, technology and the broader society. The makers of theoretical-philosophical knowledge were to be found primarily at universities that started to be established in the 11th century, as part of a rediscovery of the writings of ancient Greece and Rome that had been unread during the "dark ages." The universities took their model from Arabic precursors and much of the work in the beginning consisted of translating from Arabic to Latin the works of Aristotle and Plato and other ancient thinkers and then discussing them in learned discourse and disputation. The makers of practical, or technical knowledge, most often self-taught and unable to read and write Latin established guilds and workshops where they could "learn by doing" without much contact with the universities. And as for the makers of phronetic knowledge, the judgments of politicians and priests were based on precept and doctrine, rather than on philosophical analysis and debate.

What distinguished Europe from the rest of the world, however, was that there was an overarching Christian world-view, or cosmology with a supernatural God that would serve as a cultural seedbed for the emergence of science and technology. As with Islam, Christian theology made a fundamental distinction between humans and non-humans, between what was Godly and what was not, which would provide a congenial discursive framework for the emergence of science and technology, as well as specific sites, or cultural contexts for fostering a hybrid imagination. In the waves of monasticism, different "orders" emerged in what might be called pre-modern social and cultural movements, protesting against the way the theology was put into practice by the papal authorities in Rome.

In the monasteries of the Benedictine, and later the Bridgittine, Cistercian, Franciscan and Jesuit orders, monks carved out spaces in medieval society where they could honor their God with work and devotion and mix the different forms of knowledge into new combinations. It was Benedictine monks who were the key change agents in the agricultural revolution of the early middle ages, and it was Cistercians and Franciscans who were among the key innovators in the industrial revolution of the 12th century. Jesuits would play a major role in the scientific revolution. They were in many ways marginal in the medieval societies in which they lived, but it was nonetheless the waves of monasticism that provided, according to many historians, the crucial cultural contexts for the combination of thinking and doing, knowing and acting that would prove so crucially important – and so problematic, as well – for the development of Western civilization [Ovitt, G., 1987, Pacey, A., 1974, White, L., 1978].

Already in the 12th century the basic contours of modern, or "Western" science and technology had begun to be articulated, or at least envisioned, by monks in different ways in the different parts of Europe. In the south, where the universities first began to be established, the 12th century brought with them a rediscovery of the philosophical writings of antiquity, which would lead to a new more academic form of theology. This was a time when the works of Aristotle, translated into Latin from Arabic, were beginning be reread in Europe, and from that time, some of the first attempts were made to include the "mechanical arts" (or *techné*) in the classification of knowledge, as Aristotle had done in his *Nicomachean Ethics*, but the early Christians had not done.

Hugh of St. Victor, a monk in France, went further than most in suggesting that the mechanical arts did not merely represent a form of knowledge of their own, but that instruction in them could very well have a role to play in the university curriculum, for they taught something important about our human nature. Hughs interest in the mechanical arts would not be shared, however, by most of the school-men, or scholastic philosophers who would follow and come to be in charge of

The Cultural Story Line

o *A religion of the book, a supernatural God*

o *Separation of nature and humanity*

o *Monasticism and labor discipline*

o *The rediscovery of ancient wisdom*

o *A magical belief-system and sense of wonder*

o *The hybrid imagination of the Renaissance*

the medieval universities and the development of their curricula in the ensuing centuries. As with Aristotle, the scholastics viewed the mechanical or technical arts as non-scholarly, and it would not be until a new kind of humanist philosophy entered into the learned life-world of the university in the 15th century that science and technology would begin to be taught in a serious way. In this respect, the rediscovery of the lost mysteries of magical and alchemical knowledge and, not least, the magic of mathematics and numbers among humanist philosophers helped to open European minds to the kinds of "experimental" methods that would be so important in the scientific revolution to come.

In northern Europe, the interest in science and technology at the first universities would be somewhat different. Already in the 12th century, Roger Bacon at Oxford, who later seems to have left the university to live and work at a Franciscan monastery wrote that "I have learned more useful and excellent things without comparison from very plain people unknown to fame in letters than from all my famous teachers" (quoted in Ovitt, G., 1987, p. 119). Bacon took a great interest in the mechanical arts and apparently interacted with many of the more active inventors in the 12th century: the monks in the rapidly growing Cistercian order and his own Franciscan order that spread across Europe, bringing watermills and new agricultural techniques with them. But as in southern Europe, the forces of resistance and reaction proved too strong, and it would not be until the Reformation of the 16th century that more explicit institutions for science and technology would be established.

In central Europe, there seems to have been a closer connection between scholars and craftsmen already in medieval times. The famous treatise of Theophilus, *De diversis artibus* from the early 12th century, written by a metallurgist who is thought to have lived in a Benedictine cloister somewhere in northwestern Germany, has long fascinated historians of technology for its passionate defense of craftsmanship as a form of religious devotion, and its affirmation of an active relation to God as being part of the scientific life rather than a merely spiritual or contemplative one. Lynn White has called it a "thoughtful, theologically grounded handbook for the literate and sensitive craftsman," and it seems to have been written as a response to the denigrating attitude to craftsmen among the scholars of the time [White, L., 1978].

It is worth noting that, as with Christianity, universities came significantly later to eastern and central Europe than they did in the south and west. The first university in the area was established in Prague in 1347 (followed later in the century by Vienna, Erfurt and Heidelberg, among others), and so technological development, and with it the societal importance of the mechanical arts, had thus advanced to a significant extent from the primitive state they had been in when the first universities had been created in Italy, France and England some 250 years earlier. By the time printing with movable type was invented – by Johan Gutenberg, a learned artisan in Germany in the 1430s who apparently spent some time studying at Erfurt – a number of scholars, especially in central and eastern Europe, had begun to show a more serious interest in things technical, as part of the social and cultural movement that has come to be called the Renaissance.

In the second half of the 15th century, among the so-called artist-engineers of the Italian Renaissance, the mixing of the different types of knowledge served to inspire some of the greatest achievements that human beings have ever accomplished. They drafted and designed machines and weapons, experimented with medicines and materials, applied the ideas of optics and mechanics to their practical pursuits, in addition to producing the innumerable drawings and paintings and murals and sculptures, as well as the churches and palaces for which they are perhaps best known. Their pursuit of knowledge knew no bounds. Nor were they concerned with boundaries or compartments. Science, art, philosophy, engineering, architecture, music, as well as spiritual and mystical teachings were all drawn upon and mixed together as they sought to infuse a new kind of ambition and creative energy into Western civilization.

Leonardo da Vinci remains after 500 years the archetype, combining, in his very person, what had previously been, in Europe as elsewhere, the separate life-worlds of the learned scholar and the practical world of the craftsman [Zilsel, E., 2000]. Leonardo and some of the other so-called "Renaissance men" were able to bring those worlds together, not only because of their own personal talents and skills, which were of course substantial. They were also able to benefit from what social movement analysts call a "political opportunity structure." In the highly competitive Italian city-states, and especially in Florence, there were patrons, like the Medici family, who were willing and able to pay for their services [Grafton, A., 2000, Jardine, L., 1996].

• • • | *The Renaissance Men*

- o *Leonardo, Botticelli, Michelangelo and co.*
- o *Artists and engineers in combination*
- o *Humanism combined with magic*
- o *Leads to the invention of experimentation*
- o *A kind of collective creativity*
- o *A new vision of humanity: homo faber, man the maker*

3.3   THE RENAISSANCE AS A MOVEMENT

Since the 1960s, the Renaissance has come to be seen by many historians as an integrated and self-constructed movement firmly rooted in the social and political structures of the period. As the historian Alison Brown has noted, the Renaissance resembles the Enlightenment in that both were movements expressing new kinds of values: "Like the Enlightenment," she wrote, "it was a movement that represented a "capsule"of values and concerns regarded by its protagonists as progressive, in being based on reason and light in place of superstition and darkness"[Brown, A., 1999, p. 5].

In the words of Frances Yates,

> The word 'renaissance' means 'rebirth' and it is expressive of the way the movement was understood by the scholars and thinkers who created it. They believed themselves to be reviving, or returning to, earlier and better times, not abandoning the past for the future, but seeing the future as a child of the past. The view of time as a cyclic movement, from pure golden ages through successive worse ages of bronze and iron, encouraged the thought that the search for truth was of necessity a search for the early and the ancient, the purity of which had been lost or corrupted…The Renaissance thinkers and scholars wished to return to, or to recreate in their present, the civilization of classical

antiquity. In so doing they created new forms in art and thought and culture [Yates, F., 1972, p. 11, 12].

The Renaissance was first characterized as a distinct period or "age" in European history in 1858 by the French historian Jules Michelet and in 1860 by Jacob Burckhardt in his seminal and influential work *The Civilization of the Renaissance in Italy*. Burckhardt here defined the Renaissance as a period that embraced all aspects of Italian life at the time – political as well as social and cultural. In his book, Burckhardt saw the Renaissance as the harbinger of the Modern Age and with it as the breakthrough of modern individualism.

According to Burckhardt, the fundamental achievement of the Renaissance was that it burst what he called the "veil of ignorance and superstition" that had dominated the middle ages, and thus allowed an objective, or scientific consideration of reality to become possible. Francesco Petrarca (1304-1374) or simply Petrarch as he is called in English may be seen as the first humanist philosopher and the first author to conceive of the centuries between the fall of Rome and his own time as an age of darkness. The humanist philosophy emphasized the importance of man and his values. Thus the "dignity of man" was a favorite theme for many Renaissance humanists.

Paul Johnson gives a good description of the humanist as a social type and his institutional affiliation.

The humanists disliked, and reacted against, not only the curriculum of the universities but their reliance on the highly formalized academic technique of public debate and questions and answers to impart knowledge…Hence the humanists were outsiders, and to some extent non-academics. They associated universities with the kind of closed-shop trades unionism also found in the craft guilds. Universities in their view stamped on individualism and innovation. Humanist scholars tended to wander from one center of learning to another, picking up their choice fruits, then moving on. They set up their own little academies…but they also attached themselves to noble and princely households, which could make their own rules and were often avid to embrace cultural novelty [Johnson, P., 2002, p. 29].

Giovanni Pico Della Mirandola (1463-1494) has come down through history as the main exponent of humanist philosophy. Personally, he was an aristocrat, a brilliant intellect and a bridge builder. His philosophical approach was basically eclectic, his style rhetorically eloquent, and his aim was generally syncretic or ecumenical. He made the attempt to achieve a synthesis between Platonism and Aristotelianism – the two main philosophical traditions of ancient Greece, the one "idealistic" the other "materialistic" – and he was well versed in the Arabic and Hebrew languages and the first Western scholar who acquainted himself with Jewish Cabbala. Cabbala is a set of esoteric teachings that are meant to explain the relationship between an eternal and mysterious Creator and the mortal and finite universe (his creation). Pico was a close friend of Marsilio Ficino and a member of the Platonic Academy that Ficino had established in Florence.

Ficino was Pico's mentor and later on his colleague, and Pico most likely got some inspiration from Ficinos *Platonic Theology* in the sense that it enabled him to clarify his own highly original position. Both Ficino and Pico were concerned with God as the highest possible form of knowledge which in principle can be gained either spiritually as in mystical traditions or by philosophical contemplation. Both of them argued in favor of the philosophical way by means of mans rational powers. Pico's *Oration on the Dignity of Man (1486)* is often referred to as the manifesto of the Renaissance and is considered by many scholars a significant advance in humanist philosophy. It may be seen as a declaration of the importance of human freedom and capacity. Pico is among the first to elucidate the idea that by exercising his free will man is capable of altering the metaphysical chain of being.

We have given you, o Adam, no visage proper to yourself, nor endowment properly your own, in order that whatever place, whatever form, whatever gifts you may, with premeditation, select, these same you may have and possess through your own judgment and decision. The nature of all other creatures is defined and restricted within laws which We have laid down; you, by contrast, impeded by no such restrictions, may, by your own free will, to whose custody We have assigned you, trace for yourself the lineaments of your own nature. I have placed you at the very center of the world, so that from that vantage point you may with greater ease glance round about you on all that the world contains. We have made you a creature neither of heaven nor of earth, neither mortal nor immortal, in order that you may, as the free and proud shaper of your own being, fashion yourself in the form you may prefer. It will be in your power to descend to the lower, brutish forms of life; you will be able, through your own decision, to rise again to the superior orders whose life is divine [della Mirandola, G.P., 1486, p. 2].

The quote illustrates a transition from a God centered universe to a human centered universe – a transition which has become symbolically illustrated by Leonardo da Vinci in his drawing from 1487, the Vitruvian Man. Leonardo here portrays the harmonic proportions of man within a circle expressing the idea that man is a microcosm, or little world within the larger circle of the universe or macrocosm. Moreover, the quote also illustrates the fact that before the "fall" man is in a state of absolute potentiality where he is destined to be his own creator.

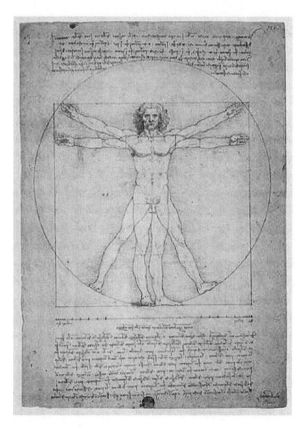

The fact that Pico ascribes metaphysical power and almost absolute transcendental capacity for human perfection to humankind has had the consequence that Pico has been reinvented in recent years by the so-called trans-humanist movement, which promotes the idea of unlimited human enhancement by use of technology, literally "playing God" (cf. Bostrom 2005). It would seem that this kind of reinvention of Pico's philosophy represents a shift from a hybrid imagination to a kind of hubris, which can be said not only to run counter to Pico's contemplative aspirations but also overemphasize the power of humans.

It is important to bear in mind that Pico never claimed that all humans were born free. The freedom of the will or simply the freedom with which Pico was concerned was neither a political, technological nor a sociological freedom. He was concerned exclusively with the metaphysical side of human freedom, and this freedom was situated in the realm of contemplation and was the exclusive possession of philosophers. The freedom of the philosopher lay in his capacity to acquire knowledge.

Leonardo da Vinci and the other Renaissance artist-engineers combined the reflective interest in theoretical-philosophical knowledge of the humanists with the practical-technical interest of the artisan world in which they lived and worked. As his notebooks bring out quite clearly, Leonardo sought knowledge not merely, or even primarily for its own sake or for purposes of contemplation, but for improving things, and in particular to make his own artistic things more brilliant and marketable. He sought an understanding of the mysteries of existence, as did the humanist philosophers. But he also needed to earn a living, and so he kept his speculations to himself, even though, as Fritiof Capra has recently suggested, his theorizing represented a more comprehensive, or holistic understanding of reality than the mechanical philosophy that would be developed by Newton and Descartes in the 17th century [Capra, F., 2007].

3.4   FROM MOVEMENTS TO INSTITUTIONS

As the cognitive praxis of the Renaissance movement spread northward, first into central Europe and then onto Elizabethan England in the course of the 16th century, it stimulated new, more "scientific" approaches to the investigation of nature, as well as new ideas about utilizing natural materials for human purposes as part of capitalist enterprise.

There was the publication of a new technical literature that depicted the techniques of mining and metallurgy, of pumps and machines, both real and imagined, and there were books, as well, that presented the procedures of various trades and skilled practices (cf. Grafton, A. and Blair, A., 1990). In the 16th century, with printing and the journeys of exploration, there was a kind of all-encompassing opening up of the European culture to new experiences and ideas, out of which modern technology and science would eventually be constituted and circumscribed.

One of the most important figures of the early 16th century was the chemist and medical doctor Philippus Aureolus Theophrastus Bombastus von Hohenheim, better known as Paracelsus, who combined a questioning of established religious and political authority with an interest in scientific observation, mathematics, mechanics and technical improvements [Grell, O.P., 1998, Pachter, H., 1961]. Paracelsus was the son of a physician in Zurich, and he attended universities, although he never managed to stay very long in any one place. He seems to have earned his living as a wandering physician and itinerant university lecturer in the towns and cities of Switzerland and southern Germany, but was continually in conflict with the authorities, and often forced to flee into the countryside. He traveled in the mining regions, where he worked as a doctor and compiled his works

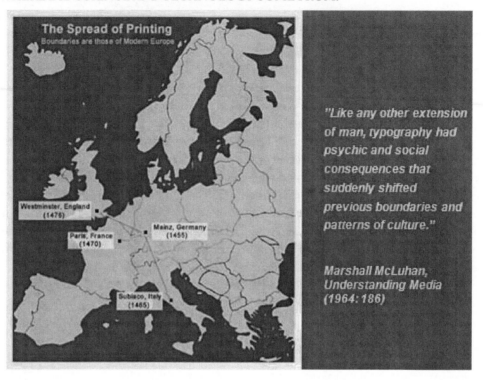

The Spread of Printing
Boundaries are those of Modern Europe

Westminster, England (1476)
Paris, France (1470)
Mainz, Germany (1455)
Subiaco, Italy (1465)

"Like any other extension of man, typography had psychic and social consequences that suddenly shifted previous boundaries and patterns of culture."

Marshall McLuhan, Understanding Media (1964: 186)

of chemical philosophy, reforming the spiritual and mystical teachings of the alchemical tradition on the basis of his own experiences. As Henry Pachter has written,

> Chemistry then was a technique rather than a science. As distinguished from alchemy, the study of the Universe, it was called "the art of distillation." Largely due to Paracelsus' efforts, this craft won recognition among the sciences. He was the first to use the word chemistry. The transition is most interesting. Neither the kitchen recipes of practical distillation nor the speculations of alchemy could lead to a scientific view of chemical processes…New conceptions were necessary before the drudgery over the chemical oven was illuminated with scientific insight [Pachter, H., 1961, pp. 97–98].

Paracelsus was one of the very first people to teach at a European university in the vernacular, and he invited practical people – apothecaries, metallurgists, miners, craftsmen – to his lectures. His approach to medicine combined the spiritual and the practical, and the theories he developed represented a kind of proto-chemistry, as he sought to provide general principles for the observed activities of metallurgists and miners, doctors and alchemists. In this sense, Paracelsus was a transition figure. In his own time, he was a rebel, opposing the medical establishment with his chemical cures and his original ideas. Seen in retrospect, however, his approach to medicine helped pave the way for modern chemistry as both a science and a huge branch of industry.

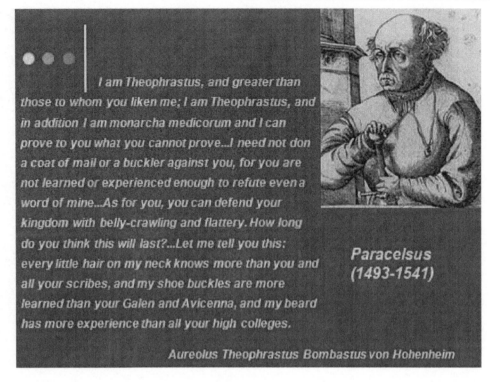

I am Theophrastus, and greater than those to whom you liken me; I am Theophrastus, and in addition I am monarcha medicorum and I can prove to you what you cannot prove...I need not don a coat of mail or a buckler against you, for you are not learned or experienced enough to refute even a word of mine...As for you, you can defend your kingdom with belly-crawling and flattery. How long do you think this will last?...Let me tell you this: every little hair on my neck knows more than you and all your scribes, and my shoe buckles are more learned than your Galen and Avicenna, and my beard has more experience than all your high colleges.

Aureolus Theophrastus Bombastus von Hohenheim

Paracelsus
(1493-1541)

The Danish nobleman Tycho Brahe was one of those who helped take the cognitive praxis of the Renaissance movement further north. As a young man studying in the new Lutheran universities in Germany, Brahe had been inspired by the teachings of Paracelsus, as well as by the new ideas about the movements of the heavenly spheres of Nicolai Copernicus, whose enormously influential book, propounding a new heliocentric theory of planetary motion was published the year after Paracelsus died. It would be Tycho Brahe and his particular kind of hybrid imagination that would play a crucial role in the long battle to win acceptance for the new cosmology. By combining the Copernican mathematically-based theory with a highly developed form of observational practice, Brahe provided the data that would then be used for the further development of the theory by Kepler, Galileo and Newton in the 17th century. Brahe was an important transition figure in mixing craftsmanship and scholarship into what would become modern science.

On his return to his native Denmark in 1577, Brahe was given the island of Hven in the straits that now separate Sweden and Denmark by the Danish king, where, for twenty years, he carried out systematic observations of the heavenly bodies as well as alchemical experiments in the spirit of Paracelsus [Christianson, J., 2000]. When royal support was withdrawn in 1597, he moved on to Prague and met Johannes Kepler shortly before he died, and he could pass on his observational data.

Tycho Brahe on Hven

Hven – or Venusia as Tycho Brahe called his island – was perhaps the most important place in 16th century Europe where there began to emerge, in an explicit and coherent manner, the modern scientific way of life. Brahe turned down a professorship at the university in Copenhagen, and he neglected his "traditional" responsibilities as a feudal lord in order to accomplish something new. At Hven, visitors could behold some of the most advanced astronomical instruments imaginable (this was before the telescope), they could wander through gardens filled with exotic plants and eat fish cultivated in specially-designed ponds, they could see buildings constructed with the aesthetic principles of Renaissance artists, they could enter laboratories designed for systematic pharmaceutical experimentation, and they could even have a look at a printing press where the scientific results were published along with Brahe's not terribly brilliant attempts at poetry.

On Tycho's island, science was given a public face, and as the years passed, hundreds of people, both craftsmen and scholars and even some politicians, from all over Europe spent time there. Hven became something of a tourist attraction, marked off on maps and praised by royalty both inside and outside of Denmark. It was also a training center for a large number of students and colleagues, who worked with Brahe on his many and varied projects. It is important to recognize that Brahe created a place where people from an aristocratic background, many of whom would perhaps have become "traditional" scholars, public servants, or priests, could work together with poorer people whom they normally would never have had the opportunity to meet.

Like Gresham College in London, where John Dee, one of the main "movement intellectuals" in what Yates, F. [1972] would term the "Rosicrucian Enlightenment" taught mathematics and number magic to craftsmen, Venusia was a site for hybrid identity formation. At Gresham, craftsmen could be exposed to philosophy and mathematics and a kind of scholarly distance to reality, and as Hill, C. [1965] has suggested, Gresham thus served to provide some of the intellectual origins of the revolution to come in the mid-17th century. A Gresham graduate was no longer just an artisan, he had something, we might say, of a hybrid imagination. It is most likely that Francis Bacon had Venusia and Gresham in mind when he wrote his vision of "New Atlantis," taking the hybrid imagination into the realm of hubris [Elzinga, A. and Jamison, A., 1984].

3.5   THE SCIENTIFIC REVOLUTION

In the 17th and 18th centuries, the discourses, institutions and practices of modern, or Western science and technology took shape in a multi-level process of cultural appropriation. There was, on the one hand, the elucidation of a new kind of rationality, or world-view, based on a new mathematics: Isaac Newton's mechanical philosophy. The book of nature was rewritten in terms of mechanical forces and laws of physical motion. In the process, nature was stripped of its mystery and turned into the raw materials for human experimentation – and eventually exploitation and domination. In Francis Bacon's enormously influential cosmology, the separation of theory and practice was transcended so that knowledge could become "useful" – and powerful.

Francis Bacon played an important role in this story not so much because he himself performed significant experiments but because he envisioned the emergent character of experimentation. Three features characterize the Baconian program for the advancement of science and technology:

An awareness of the importance of appropriate research procedures (the scientific method),

A clear vision of the purpose of the scientific enterprise (improving the human condition), and

A practical understanding of the arrangements necessary to put the program in practice (scientific institutions and state support).

The imagery of nature related to the unfolding of Bacon's program marked a transition from an organic to a mechanical conception of nature and a shift in the meaning of theory. Theory had previously meant contemplation of eternal cosmic order and beauty. In contrast, the meaning of theory in the Baconian program is that of an instrument for prediction and control. Theory enters into practice in the form of an engineering attitude towards the world. This attitude and not the technological artifacts in themselves is what the German philosopher Martin Heidegger in *The Question Concerning Technology* from 1954 sees as the essence of technology. Harnessing a disenchanted nature for human exploitation is what he calls "enframing" with a romantic undertone of cultural despair. From a nurturing mother nature has come to be merely raw material. In philosophical

Modern Science as Cultural Appropriation

● ● ●

o *At the discursive level:*
- *A semantics of utility: "useful knowledge"*
  - *A grammar of mathematics*
    - *A language of mechanics*

o *At the institutional level:*
- *Media of communication*
  - *Academic organizations*
    - *Professional norms and quality standards*

o *At the practical level:*
- *Hybrid identities*
  - *Technical applications*
    - *Experiments, instruments and methods*

aesthetics, nature would later be singled out as a place for contemplating the sublime and the beautiful or as an escape from industrial civilization.

In his *Wisdom of the Ancients* from 1609, Bacon was fully aware of the importance of imagery and myth as being far more than mere rhetorical decoration. Through imagery and myth, Bacon states that "inventions that are new and abstruse and remote from vulgar opinions may find an easier passage to the understanding." Especially his interpretation of the ancient myth of Prometheus is a good illustration of the new relationship between man and nature and the role of experiment. Bacon posits that "…the nature of things betrays itself more readily under the vexation of art than in its natural freedom."

What eventually came to be characterized as modern science represented a form of knowledge production that drew much of its inspiration from earlier and much broader social and political struggles, and most especially from the Puritans and other religious dissenting groups of the mid-17th century. It had been a more all-encompassing project: to "turn the world upside down" as one pamphleteer of the English Civil War put it: "to set that in the bottom which others make the top of the building, and to set that upon the roof which others lay for a foundation" (quoted in Hill, C., 1975, p. 13).

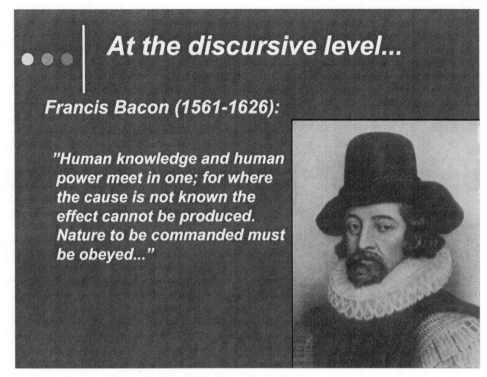

At the discursive level...

Francis Bacon (1561-1626):

"Human knowledge and human power meet in one; for where the cause is not known the effect cannot be produced. Nature to be commanded must be obeyed..."

One of the pamphlets from that period which has had a lasting impact on history was *The Law of Freedom in a Platform*, published in 1652, by Gerrard Winstanley of the "Diggers," in which he articulated a visionary program for an egalitarian way of life, which included some rather far-reaching ideas about science and technology. For Winstanley, the making of knowledge was to be carried out collectively, without any thought of profit or personal gain. The "Government under Kings," Winstanley wrote, had been based on "Traditional Knowledge and Learning" by which "both Clergy and Lawyer...by their cunning insinuations live meerly upon the labor of other men." In the "Common-wealth" to come, it would be different.

In the process which led to what has come to be called the scientific revolution, there was an explicit narrowing of focus, a reduction of ends to means in the consolidation of modern scientific and technological rationality. There was a gradual sharpening of tools, both conceptual and practical, with the formation of concepts, methods and instruments, and there was a formalization of the organizational activities, which came with the establishment of key institutions in the 1660s, such as the Royal Society and the French Academy of Sciences. The social contract that was established in the 17th century, between the new-fangled "experimental philosophers" and their royal patrons, systematically excluded many of those who had previously wanted to be included in the reform of philosophy – women, in particular. In order to become acceptable to those in power, the scientists

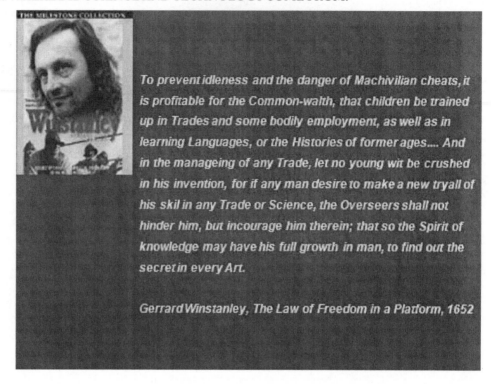

To prevent idleness and the danger of Machivilian cheats, it is profitable for the Common-walth, that children be trained up in Trades and some bodily employment, as well as in learning Languages, or the Histories of former ages.... And in the manageing of any Trade, let no young wit be crushed in his invention, for if any man desire to make a new tryall of his skil in any Trade or Science, the Overseers shall not hinder him, but incourage him therein; that so the Spirit of knowledge may have his full growth in man, to find out the secret in every Art.

Gerrard Winstanley, The Law of Freedom in a Platform, 1652

needed to renounce many of their earlier political ambitions. In the words of Lewis Mumford, "Under the new ethic that developed, science's only form of social responsibility was to science itself: to observe its canons of proof, to preserve its integrity and autonomy, and to constantly expand its domain"[Mumford, L., 1970, p. 115].

Gerrard Winstanley's ideal of knowledge, like that of Paracelsus in the 16th century, was eminently practical, but it was also moral and spiritual. To Bacon's notion of useful knowledge, Winstanley and the other puritan radicals added what might be termed a sense of accountability, or social responsibility, as well as a broader meaning for the pursuit of knowledge, which has relevance for the science and technology of our own time. In the words of Charles Webster, who produced a major study of those movements in the 1970s:

> Environmental circumstances have necessitated reference to an idea of the social accountability of science, analogous to the view which the Puritans more readily derived from their religious convictions...Although the Puritans looked forward to an unprecedented expansion in human knowledge, they realized that it would be necessary to exercise stringent discipline to prevent this knowledge resulting in moral corruption and social exploitation [Webster, C., 1975, pp. 517–518].

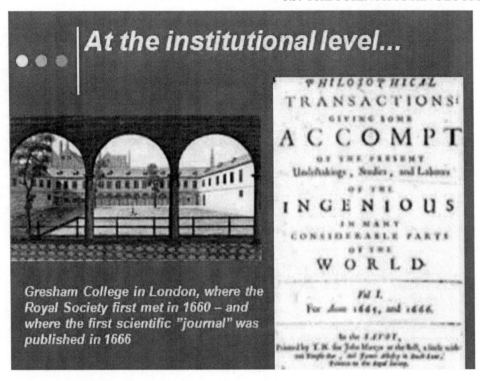

Gresham College in London, where the Royal Society first met in 1660 – and where the first scientific "journal" was published in 1666

It would be in the 1660s, directly after the restoration of the English king that science would begin to find its characteristic institutional form, and its adherents would vow not to meddle "with Divinity, Metaphysics, Moralls, Politicks, Grammar, Rhetorick or Logick," as it was expressed in the charter for the Royal Society. It was a particular form, or mode, of knowledge making which came to be institutionalized. A way of knowing reality was constituted that was technically-oriented and subservient to the interests of the powerful.

This was accomplished both through using scientific instruments and experimental rituals, as well as through the mutual interaction between scientific inquiry and technical improvement, such as took place in the development of steam power, ship-building and navigational techniques. The hybrid entities that were being constructed called for the formation of hybrid identities on the part of those who were doing the constructing. As the telescope and microscope literally disclosed new dimensions of reality, the clock and the compass provided scientists and merchants with images and metaphors which – through a kind of incorporation process – enabled them to re-cognize the natural world [Jardine, L., 1999].

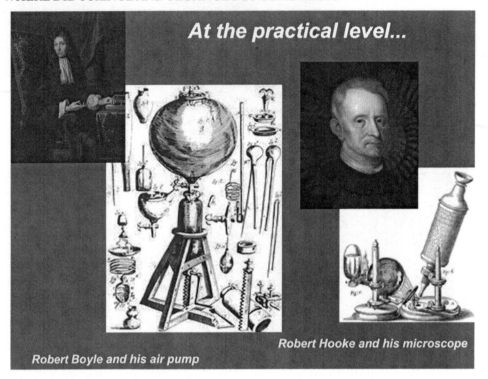

At the practical level...

Robert Hooke and his microscope

Robert Boyle and his air pump

3.6   THE ENLIGHTENMENT AS A MOVEMENT

The new discourses, institutions and practices of modern science did not go unchallenged. The experimental philosophy was opposed both by the upholders of traditional, religious knowledge, as well as by the marginalized radicals and their descendants. In the later seventeenth and early eighteenth centuries, the scientific "aristocracy" that had emerged in London and Paris at the Royal Society and the *Académie des Sciences* in Paris was challenged by new kinds of dissenting groups, as well as by representatives of the emerging middle classes. In this sense, the enlightenment was also a kind of social and cultural movement. Many of the radical dissenters fled from Europe to the colonies in North America, and some of them who stayed behind established scientific societies, often in provincial areas in opposition to the established science of the capital cities.

Many of the participants in the enlightenment movement shared with the academicians and their royal patrons a belief in what Max Weber termed the protestant ethic – that is, an interest in the value of hard work and the virtue of making money – and most had an interest in what Francis Bacon had termed useful knowledge. But the movements that inspired the French, American, and the British industrial revolutions objected to the limited ways in which the Royal Society and the Parisian Academy had organized the scientific spirit and institutionalized the new methods and theories of the experimental philosophy. There was also a geographical diffusion of the modern

The Age of Enlightenment

Voltaire

Benjamin Franklin

Rousseau

Ludvig Holberg

Adam Smith

scientific spirit to places like Sweden and Denmark and North America, where academies of science were created in the 18th century. In Scotland and Birmingham, philosophers and craftsmen met together in places like the Lunar Society and inspired the industrial revolution with their inventions and their ideas about how those inventions could be used. It was in such places that the technician James Watt and the philosopher Adam Smith could combine theory and practice in what would grow into the technological and economic sciences of the 19th century [Uglow, J., 2002].

The American Philosophical Society for the Promotion of Useful Knowledge, founded in Philadelphia in 1768, was a center of scientific and technical activity in the colonies as well as after independence. Under the leadership of Benjamin Franklin, it was a place where the particularly American form of "republican" science and technology was fostered. Franklin himself was an archetypical hybrid, taking active part both in the political life of the United States by serving as a diplomat and public servant, as well as its intellectual, and economic life, as an inventor, scientist and businessman.

In France, the writings of Voltaire and Rousseau, mixing satire and critical commentary into their visions of human progress and creativity played an important role in preparing the way for the French Revolution. Particularly important was the long-term effort by Denis Diderot to bring scholars and craftsmen together in producing the articles for his *Encyclopedie*. The various attempts to democratize scientific and technical education in the wake of the French Revolution and to apply the mechanical philosophy to social processes – i.e., to view society itself as a topic for scientific research and analysis – indicate how new forms of scientific and technological practice and new institutions are inspired by broader movements. The revolutionary government was the first to establish a science based institution of higher education, the *École Polytechnique*, and it was there, as we shall see in the next chapter, that visions of a technocratic order developed in the writings of the count Saint Simon, and his secretary, August Comte. The institutionalization process included the articulation of new philosophies of science – positivism, in particular – and new disciplines, related to the emergent needs of an industrializing social order: statistics, geology, thermodynamics, political economy, and, most especially perhaps the various forms of "applied" or technical sciences that would come to be called technology. Out of social and cultural movements, there emerged new scientific institutions and disciplines, new forms of knowledge making.

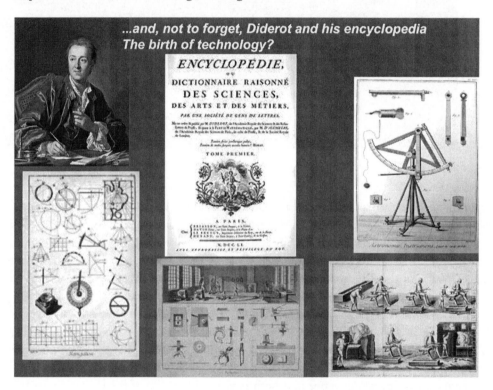

CHAPTER 4

Science, Technology and Industrialization

The true man of science will know nature better by his finer organization; he will smell, taste, see, hear, feel better than other men. His will be a deeper and finer experience. We do not learn by inference and deduction, and the application of mathematics to philosophy, but by direct intercourse and sympathy. It is with science as with ethics – we cannot know truth by contrivance and method; the Baconian is as false as any other, and with all the helps of machinery and the arts, the most scientific will still be the healthiest and friendliest man, and possess a more perfect Indian wisdom.

Henry David Thoreau, "Natural History of Massachusetts" (1842)

4.1 INDUSTRIALIZATION AS A CULTURAL PROCESS

In the late 18th century, something happened that has come to be called the "industrial revolution." The term became popular in the 1880s after it was used in a series of lectures by the historian Arnold Toynbee to characterize economic development in England from 1760 onwards. By so doing, Toynbee implied that the economic changes that had taken place in England were comparable to the political changes that had taken place in France in the late 18th century: hence the use of the term revolution to describe them. But as with the scientific revolution of the 17th century, there has been little agreement ever since among historians about how and why the industrial revolution took place, or, for that matter, what industrialization was all about.

The economically-minded tend to see it as a mechanization process by which the infusion of machines and the mechanical philosophy with which they were associated turned the economy into something fundamentally different than it had been before. The socially-minded have tended to emphasize the "social innovations" that characterized industrialization, in particular, the factory system and the new kinds of governmental institutions and social organizations that emerged in the course of the 19th century to shape an industrial society. In this chapter, we try to show how both the economic and social changes that are associated with industrialization can be seen as part of a broader cultural transformation, a change in attitudes and behavior from rural communities to an industrial way of life.

What was/is Industrialization?

- **An economic and technical revolution**
  - *the growth of modern industry, or mechanization*

- **A process of social change**
  - *the coming of industrial society and its institutions*

- **Cultural, or human transformations**
  - *from rural communities to an industrial way of life*

As we noted in chapter two, the most important foundational narrative for the economic story-line was provided by the philosopher-turned-political economist Karl Marx and his long-time friend and collaborator Friedrich Engels, who identified technological development as the main factor behind the growth of what they termed modern industry. Already in 1845, in *The Condition of the Working Class in England*, Engels had written of how the technical inventions of the late 18th century "gave the impulse to an industrial revolution, a revolution which at the same time changed the whole of civil society" (Engels, quoted in Williams, R., 1977, p. 138). A few years later, when Marx and Engels in 1848, wrote the *Manifesto of the Community Party*, a document that would come to play a fundamental role in the making of socialist movements, they had come to characterize industrialization as a kind of permanent revolution, based on the changes in the means of production brought about by technology:

> All that is solid melts into air, all that is holy is profaned, and man is at last compelled to face with sober senses, his real conditions of life, and his relations with his kind... Subjection of Nature's forces to man, machinery, application of chemistry to industry and agriculture, steam-navigation, railways, electric telegraphs, clearing of whole continents

for cultivation, canalisation of rivers, whole populations conjured out of the ground –
what earlier century had even a presentiment that such productive forces slumbered in
the lap of social labour? [Marx, K. and Engels, F., 1968, p. 38, 40].

As Marx later put, in his major work, *Das Kapital*: "by means of machinery, chemical processes and other methods, it [modern industry] is continually transforming not only the technical
basis of production but also the functions of the worker and the social combinations of the labor
process"[Marx, K., 1976, p. 617].

In the 1930s, the Austrian Joseph Schumpeter added to the story-line with his notion of
"creative destruction." He identified three "waves" of industrial development that had been insti-
gated by clusters of what he termed radical innovations, which had led to far-reaching repercussions
throughout the economy and the creation of new branches of industry and the destruction of older
ones. In the 1980s, Christopher Freeman and other students of innovation identified a fourth wave
in the mid-20th century, based, among other things, on petrochemical products, mass automobil-
ity and atomic energy that had begun its decline in the 1970s, but which was beginning to be
"creatively destroyed" by a new wave based on innovations in information technologies, such as
personal computers, video recorders, and software systems and biotechnologies, especially genetic
modification [Freeman, C., 1987].

In this and the following chapters, we will focus on the role that cultural and social movements
– and people with hybrid imaginations – have played in these processes. For it is our contention that it
has been in the periods "between the waves" when the economic potential of the radical innovations
has begun to fade that industrial development has not only been creatively destroyed but reimagined
and eventually reconstructed, as well. As such, what we term cycles of creative reconstruction can be
thought of as the interaction between social and cultural movements and technological and scientific
developments. By responding to the experiences of one wave, cultural and social movements have
served to reimagine and reconstruct the next.

In the course of the 19th century, there were two waves of industrial development that were
followed by two periods of active mobilization in social and cultural movements. In the 1810s and
1820s, a broad-based cultural movement, which has come to be called romanticism, emerged among
artists and writers and, as with the more politically-oriented cooperative movement, had a major
influence on many scientists and engineers. Later, in the 1860s and 1870s, socialist and populist
movements, combining culture and politics, also served to foster hybrid imaginations that would
affect the reconstruction of science and technology.

But as they contributed to the reconstruction of science and technology in the ensuing waves
of industrial growth and expansion, something happened to the ideas and practices of the social and
cultural movements. Those who would keep the spirit of the movements alive waged a struggle on two
fronts, contending, on the one side, with the incorporation pressures of the dominant commercial
culture, and, with it, the discourses of positivism and imperialism, and on the other side, with
the accommodation strategies of residual national cultures and nationalist ideologies. As such, the

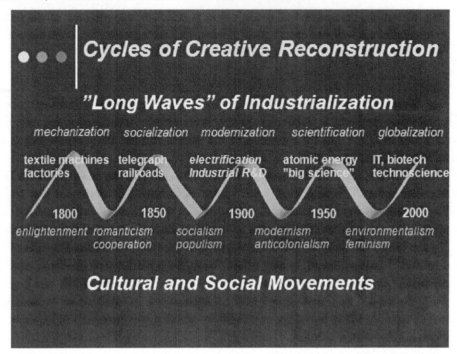

cultural appropriation of science and technology in the course of industrialization was a process of contending response strategies on the part of the three cultures of science and engineering.

4.2   THE FIRST CYCLE

One of the most important differences between the first and second cycles, or waves of industrialization, concerns the role played by science and scientific knowledge. In the first cycle, the radical innovations themselves – textile machines, coal mining and processing technology, improvements in the steam engine - were not directly based on science. In the words of Samuel Lilley,

> The early stages of the Industrial Revolution – roughly up to 1800 – were based very largely on using medieval techniques and on extending these to their limits… This is not a story of sophisticated inventions breaking through some critical technological barrier, and so creating the conditions for expansion. Developments that were technically so simple can only be responses to social and economic conditions that offered widening opportunities for self-advancement through innovation [Lilley, S., 1975, p. 190, 195].

What science provided was a new belief system, a change of attitude that would have far-reaching effects in the subsequent cycles. In the words of Alfred North Whitehead [Whitehead, A.N., 1925, p. 98], "the greatest invention of the nineteenth century was the

The First Cycle

- *"the industrial revolution" (ca 1780-1830)*
- *iron, textile machines, and steam engines*
- *technologies of mechanization*
- *the factory as an organizational innovation*
- *social and cultural movements:*
  - *"machine-storming" and romanticism*
  - *cooperation and polytechnics*

invention of the method of invention. A new method entered into life." As Lilley, S. [1975, p. 226] puts it,

> During the nineteenth century the balance between technological innovations and economic incentives changed radically. New technologies arose which were not in any important degree extensions of previous ones (as the spinning machinery had been) or substitutes for resources in short supply (coke-smelting, railways). They arose very nearly as if invention were regarded (by the inventor and by society) as a good in itself.

Before industrialization, most people lived in the countryside, and they obtained their livelihoods from agriculture, animal husbandry, fishing, and forestry, and for the most part, knowledge making was directly related to everyday life experiences. Most of what one knew was not formalized or theoretical but rather informal and practical. Knowledge was largely produced and communicated in local settings, in processes of "situated learning" in what Etienne Wenger has termed "communities of practice" rather than in formal research and educational institutions.

The emergence of modern science in the 17th century challenged these traditional forms of knowledge-making. Science provided ways of knowing – mathematical logic and calculus, systematic experimentation, mechanical models, chemical theories – that separated knowledge from everyday life. Scientific knowledge was abstract and codified, and it was communicated in writing. By means of systematic experimentation, for instance, the traditional techniques of energy production, mining and metallurgy could be made much more efficient. Using quantitative methods and mechanical models could provide much greater precision and, not least, possibilities for control and management over processes of both primary and secondary production. It would not be until the mid-19th century that scientific knowledge would be used directly in industry. More important in the early period than any particular uses of science was the change in attitude.

As Humphrey Jennings put it in his remarkable collection of documents, entitled *Pandaemonium*, tracing the "coming of the machine," the new scientific attitude, or what he calls a "fundamental alteration of 'vision'" in the course of the 18th century, was "being achieved not merely as a *result* of changing means of production but *also* making them possible" [Jennings, H., 1987, p. 38]. And the "dual revolution" – the political revolutions that ushered in democracy and the technical and economic revolutions that ushered in industrialization – were the immediate outcome, we might say, of this altered vision (cf. Hobsbawm, E., 1962).

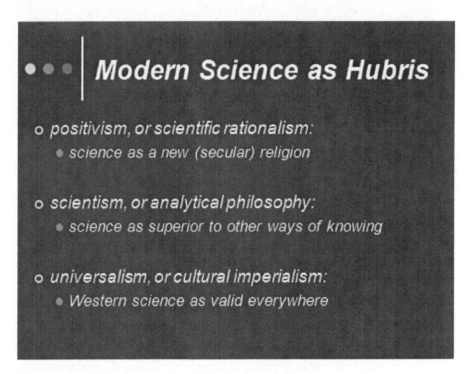

Across Europe and North America, as science gained more adherents in the course of the 18th century in the Age of Enlightenment, ever more people began to envision another kind of society, based on the applications of scientific rationality. Science, we might say, helped to open up the realm of the imagination – sometimes to such a degree that it took on the character of hubris: the over-extension into what came to be termed positivism, or the positive philosophy. First formulated by the Frenchman August Comte at the *Ecole Polytechnique*, the world's first technological university that had been founded in the 1790s by the revolutionary government, positivism turned science into a new kind of secular religion, seeing scientific rationalism as bringing on a new stage in history replacing theology and metaphysics.

In the course of the 19th century and on into the 20th, positivism would serve as an important philosophy of science, which would justify, among many scientists and engineers, an attitude that has been termed "scientism," a belief that scientific knowledge, because of the precision of its methods, was superior to all other forms of knowledge. In the early 20th century, philosophy itself, particularly in Britain, would strive to become "scientific" in one or another form of analytical philosophy. Rather than grapple with the big questions that philosophers had thought about through the centuries, analytical philosophy subjected the complexities of human life to mathematical logic, arguing, as Comte had done, that metaphysics and theology were far too speculative and un-scientific in their approach to philosophy. Both positivism and scientism would give rise to the more general idea of universalism that would be characterized by Robert Merton in the 1940s as one of the norms of science, the idea that was so widespread in the age of imperialism that western science and technology were the only valid or legitimate ways of making knowledge [Jamison, A., 1994]. And much as movements of anti-colonialism and anti-imperialism emerged in the 20th century to fight back and tame the hubris, in the 19th century a wide-ranging social and cultural movement – romanticism – emerged to humanize the machine and the process of mechanization more generally.

4.3   THE ROMANTIC MOVEMENT

The machines and factories which characterized the first period of industrialization were rejected outright by many influential poets and writers, as well as by the so-called Luddites who attacked what the poet William Blake had called the "dark Satanic Mills" of the new industrial cities in his famous poem, *Jerusalem.*

In the 1810s, workers and craftsmen, in the name of their mythical leader, Ned Ludd, systematically destroyed machines for spinning cotton yarn and weaving cloth in factories in the new industrial towns of north England. They brought out the British army, which eventually put their protests to an end, but they helped inspire a more general questioning of the industrial revolution that would become a significant cultural and social movement in the 1820s and 1830s. In the first speech that he gave to the House of Lords in 1812, Lord Byron, the famous poet, said of the Luddites:

> During the short time I recently passed in Nottinghamshire, not twelve hours elapsed without some fresh act of violence; and on the day I left the county I was informed that forty frames had been broken the preceding evening, as usual, without resistance and without detection… Considerable injury has been done to the proprietors of the improved frames. These machines were to them an advantage, inasmuch as they superseded the necessity of employing a number of workmen, who were left in consequence to starve… The rejected workmen, in the blindness of their ignorance, instead of rejoicing at these improvements in arts so beneficial to mankind, conceived themselves to be sacrificed to improvements in mechanism. In the foolishness of their hearts they

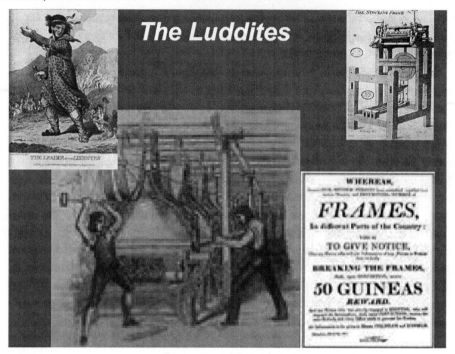

imagined that the maintenance and well-doing of the industrious poor were objects of greater importance than the enrichment of a few individuals by any improvement, in the implements of trade, which threw the workmen out of employment (Byron, quoted in Jennings, H., 1987, pp. 131–132).

Byron, along with William Wordsworth, John Keats, Percy Shelley and many other poets and artists would subsequently transform the Luddite protests into art, providing an alternative to the emerging mechanical world with another kind of knowledge making, countering the industrial revolution by inventing new forms of cultural expression. Most significant, in the long run, perhaps would be the book first published by Mary Shelley in 1818, in which the industrial or mechanical world view was embodied in "the modern Prometheus," Doctor Victor Frankenstein.

To indicate the plot of the novel, the following two quotes by the main characters, Victor Frankenstein and his creation (simply called the being), respectively, might be helpful. First, the words of the doctor:

It was on a dreary night of November, that I beheld the accomplishment of my toils... I had worked hard for nearly two years, for the sole purpose of infusing life into an inanimate body. For this I had deprived myself of rest and health. I had desired it with an ardour that far exceeded moderation; but now that I had finished, the beauty of the dream vanished, and breathless horror and disgust filled my heart. Unable to endure the aspect of the being I had created, I rushed out of the room [Shelley, M., 1994, p. 38, 39].

And then the words of his "monster" after Frankenstein decides to destroy his apparatus and refuses to make a companion for his creation:

Slave, I before reasoned with you, but you have proved yourself unworthy of my conde-scension. Remember that I have power; you believe yourself miserable, but I can make you so wretched that the light of day will be hateful to you. You are my creator, but I am your master; - obey! ...Beware; for I am fearless, and therefore powerful. I will watch with the wiliness of a snake, that I may sting with its venom. Man, you shall repent of the injuries you inflict [Shelley, M., 1994, p. 140].

Briefly put, the plot of Mary Shelley's novel can be summarized this way: "Frankenstein – a man, a scientist, creates a living being out of bits of corpses which, grown monstrous, turns on its creator and runs amok"[Hammond, K., 2004, p. 184]. During the past 200 years, the Frankenstein story about creation of life has continually been retold to remind us of the risks and dangers of science and technology, of manipulating nature and "playing God." In a straightforward interpretation, Victor Frankenstein is a clear cut example of hubris and irresponsibility. Perhaps more than any other work, Mary Shelley's novel has come to symbolize romantic attitudes about technology. Today, nearly two hundred years after its publication, it is still not outdated and continues to be read, dramatized, and filmed as a warning about the arrogance of scientists and engineers.

As the movement developed and moved into the cultural mainstream, especially in Britain, romantic writers provided a cultural assessment of industrialization, focusing attention on the mech-anization of humanity that had accompanied the mechanization of production. Most influentially perhaps in the popular novels of Charles Dickens, the cultural implications of industrial society were subjected to what might be termed the mobilization of the imagination. In his novel *Hard Times*, published in 1854, Dickens begins his depiction of the new industrial way of life by introducing his readers to the mindset of a factory owner, Thomas Gradgrind:

"Now, what I want is, Facts. Teach these boys and girls nothing but Facts. Facts alone are wanted in life. Plant nothing else, and root out everything else. You can only form the minds of reasoning animals upon facts: nothing else will ever be of any service to them. This is the principle on which I bring up my own children, and this is the principle on which I bring up these children. Stick to Facts, Sir."... In such terms Mr. Gradgrind always mentally presented himself whether to his private circle of acquaintance or to the public in general [Dickens, C., 2000, p. 3, 4].

This kind of positivist thinking was, for Dickens and other romantic writers, a form of hubris. The social acceptance of scientific ways of understanding and attempts to act rationally on these understandings led to the exclusion of other ways of knowing, both scientific and traditional. In the novel, Dickens tries to show what the exclusion of other ways of knowing may lead to. In *Hard Times*,

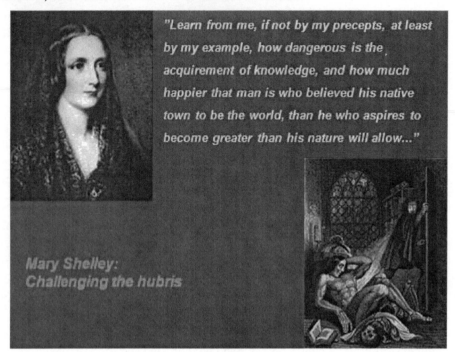

"Learn from me, if not by my precepts, at least by my example, how dangerous is the acquirement of knowledge, and how much happier that man is who believed his native town to be the world, than he who aspires to become greater than his nature will allow..."

Mary Shelley:
Challenging the hubris

the main attacks of Dickens' romantic critique of the modern project are the following: the Science of Political Economy, and the Philosophy of Utilitarianism. However, Dickens' critique is far more wide-ranging. In addition to economic and philosophical theories, it also embraces industrialism, capitalism, and education. Dickens' point of departure for his critique is Coketown – a fictional and ugly British industrial town. The depiction of Coketown presents us with a full-fledged picture of the mid nineteenth century industrial way of life: the factory at its core, the industrial capitalist in charge, the working men, the school, family, marriage, divorce, laws, class, and a subservient Parliament.

The positive values in the novel are represented by the circus, which satisfies the need for imagination and entertainment but also the embodiment of altruistic behavior where human care and affection come naturally. Dickens thus offers a powerful picture of two opposed ways of seeing and thinking about the world. The first way of seeing and thinking about the world is the dominant picture of a materialist, statistical, deterministic, theory-blinded positivist way of thinking represented by factory owner Josiah Bounderby, and hardware manufacturer, founder of a model school and later MP for Coketown, Thomas Gradgrind. The alternative way of thinking is based on idealism and the power of affection, sensibility and solidarity represented by the members of the circus, in particular the daughter of the circus owner, Sissy Jupe.

The romantic "revolt of the senses," as Whitehead, A.N. [1925] once called it was much more than merely a critique of hubris; it was a wide-ranging social and cultural movement that included new forms of knowledge making in both science and technology. In 1817, the industrialist Robert Owen started to develop his programs for factory and eventually social reform, and in the 1820s, the mechanics' institutes began to be established in order to educate workers in the new kinds of skills and knowledge that industrialization required. Among the followers of Owen, as well as in some other branches of what came to be called cooperative and Chartist movements, alternative forms of production and manufacture were developed. In some of the "utopian communities" that Owenites and others established in North America, there was a widespread experimentation activity, which generated a literature about technology as well as "appropriate" applications of machines to agriculture, transportation, and communications [Kingston, J., 1976].

The romantic movement emerged as what might be called the other side of the Enlightenment in works such as Johan Wolfgang von Goethe's *The Sorrows of the Young Werther* in 1774 and Jean Jacques Rousseau's *Reveries of a Solitary Walker* in 1782. In these works, a new romantic sensibility entered European culture. According to Löwy, M. and Sayre, R. [2001], the romantic movement was based on a critique of five main features of modern science and technology:

The disenchantment of the world.

The quantification of the world.

The mechanization of the world.

Rationalist abstraction.

The dissolution of social bonds.

The "disenchantment of the world," which would later be a key phrase in the sociology of Max Weber, means the banishing of superstition, myth, and magic, and their substitution for a *realistic* approach to the world [Greisman, H.C., 1976, p. 496]. In positivism, this new *realistic* approach to the world means that values and facts become separated. The result is a loss of meaning and existential orientation. The phenomena and processes of the world simply "are" and "happen" but no longer signify anything. As argued by Elzinga:

> Romanticism denies the basic assumptions of scientism and challenges them. Indeed, in some cases the very idea of progress in and through science is denied. Being a child of a reaction to the technocratic optimism of an earlier century, it promotes intuition as superior, while scientific reason is equated with an advance of mechanization of human life [Elzinga, A., 2002, p. 1].

4.4   ROMANTIC SCIENCE AND TECHNOLOGY

As Richard Holmes has recently shown in his book, *The Age of Wonder*, many scientists and engineers in the early 19th century were influenced by the romantic movement and what might be termed its polytechnical cognitive praxis. Combining a belief in scientific and technical progress borne out of the Enlightenment with an emotional and "hands-on" approach to making knowledge, biologists like Joseph Banks and chemists like Humphrey Davy helped humanity to discover what Holmes terms the "beauty and terror of science"[Holmes, R., 2008]. When Banks became the president of the Royal Society, he expressed his scientific ambitions in these distinctively romantic terms:

> I have taken the Lizard, an Animal said to be endowed by Nature with an instinctive love of Mankind, as my Device, & have caused it to be engraved on my Seal, as a perpetual Remembrance that a man is never so well employed, as when he is labouring for the advantage of the Public; without the Expectation, the Hope or even a Wish to derive advantage of any kind from the result of his Exertions (quoted in Holmes, R., 2008, p. 55).

As a movement romanticism was largely overtaken, or at least made problematic, by the course that scientific and technological development was to take in the second half of the 19th century, becoming more professionalized on the one hand and commercially-oriented on the other. With the coming of the industrial society, a range of new opportunities in industry and other contexts emerged, and the romantic ideals of gentlemen scientists like Banks became ever harder to live up to. For the social critics, it became ever more difficult to escape from, or oppose, the dominant ways in which science and engineering came to be organized and institutionalized. But some of the ideas of the movement did have impacts on the emerging industrial order and its ways of life, not least in regard to developing more reflective attitudes and responsible approaches toward science and technology.

As a social and cultural movement, romanticism can be said to have opened up a space in society for fostering a hybrid imagination. In Denmark, some of the criticisms that were mounted against positivism can easily be found either individually or in combination in writings from the romantic era of the 19th century. One outstanding example is the scientist Hans Christian Ørsted (1777-1851) who discovered electromagnetism and founded Polytechnic Institute in Copenhagen 1829, now Technical University of Denmark. In particular, in his book, *The Spirit in Nature*, from 1850, there is a strong influence of the romantic movement. The title of the book is a reflection of Schelling's philosophy of nature. In the romantic era of the 19th century, natural philosophy was introduced into science and technology, and it was formative in determining the proper relationship between engineers and nature, engineers and society and engineers and humanity.

In many ways, Ørsted is a good example of a hybrid imagination in action. He was well versed in both the humanities and the sciences and could be described as a member of both of the two cultures, the scientific and the literary, that, as we shall see, were influentially characterized by C.P. Snow in 1959 as the main dividing line between intellectuals. He was also both a witness and a driving force in the development of the natural sciences, engineering, and engineering education as well as philosophy and politics in Denmark [Jessen, E., 2007].

Erland U. Jessen gives a good description of Ørsted's relation to the romantic movement:

> Though Ørsted disliked the romantic worship of the genius and the excesses of many of his contemporary scientists, poets and artists, for half a century his thinking and

The Hybrid Imagination:
Hans Christian Ørsted (1777-1851)

• mixed Naturphilosophie with experimentation
• to look for the "spirit in nature"...
• discovered electromagnetism (1820)
• and founded DTU in 1829

endeavours were in accordance with Schelling's ideas of total nature and his concept of a mutual task of reconciliation. One of his problems as a scientist was that an interpretation of nature was not necessarily followed by a conviction of the unity of existence. His firm belief was that nature is one realm of reason. It is not only a material substance, but is governed by a principle of the spirit or the soul... The senses can unite and determine phenomena of nature, but are often defective or incomplete. They cannot catch the causality of events without a metaphysical reason that regulates forms and systematizes the material of the senses [Jessen, E., 2007, p. 47].

One of the people who was directly influenced by Ørsted, in particular his discovery of electromagnetism, was the painter turned inventor, Samuel Morse, whose inventions of the telegraph machine and the Morse code would be so crucially significant in the next wave of industrialization. Morse studied both art and chemistry and was in Europe trying to pursue his artistic career when he heard of Ørsted's discovery of electromagnetism. Perhaps because he was an artist as well as a scientist, Morse could appreciate the significance of the discovery more than other people. In any case, he was able to build on the discovery and come up with the solution to the problem of communicating over long distances that had previously seemed to be a "mission impossible." It took him many years to convince others that his inventions could actually be useful; like many an inventor before him, he needed to combine his inventiveness with entrepreneurship and take his hybrid imagination to the commercial marketplace before he could spawn the telegraph industry. As such, Morse is a good example of the transformation of hybridity into hubris.

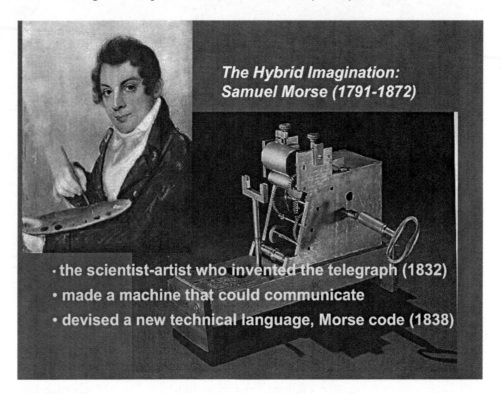

The Hybrid Imagination:
Samuel Morse (1791-1872)

· the scientist-artist who invented the telegraph (1832)
· made a machine that could communicate
· devised a new technical language, Morse code (1838)

In Denmark, a very different kind of romanticism was articulated by the priest and historian N. F. S. Grundtvig, who debated with Ørsted about the kinds of responses to industrialization that were most appropriate. In addition to writing hundreds of psalms and giving literally thousands of sermons, and writing a large number of historical and theological works, Grundtvig also played an important political role in 19th century Denmark, serving in the Parliament and, perhaps most significantly, initiating a new form of education. Grundtvig wanted to mobilize the traditions of Danish history in order to make industrialization compatible with the human values that he and many other Danes found important.

One of the more enduring legacies of his multifaceted hybrid imagination was the establishment of the people's, or folk high schools in the countryside to provide an "education for life," as he called it, rather than the education for death that was provided to the universities. At these schools, that from the mid-19th century, spread from Denmark onto the rest of Scandinavia and then further into other parts of the world, Danish farmers could learn about new agricultural techniques, as well as about new cooperative forms of organizing their food-processing business, in the cooperative slaughterhouses and dairies, which would become the main source of Danish industrialization [Borish, S., 1991].

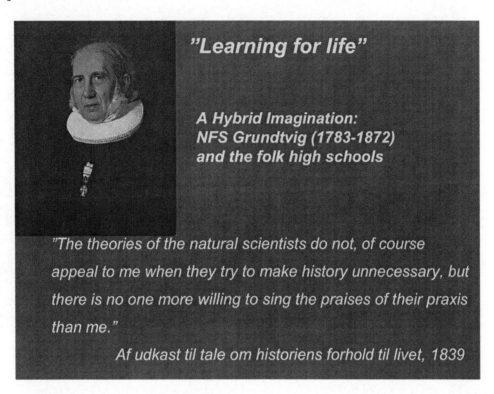

"Learning for life"

A Hybrid Imagination:
NFS Grundtvig (1783-1872)
and the folk high schools

"The theories of the natural scientists do not, of course appeal to me when they try to make history unnecessary, but there is no one more willing to sing the praises of their praxis than me."

Af udkast til tale om historiens forhold til livet, 1839

One of the scientists in the mid-19th century who was most active in keeping the spirit of the romantic movement alive was Henry David Thoreau. Withdrawing in self-imposed isolation to a house that he built himself on the shores of Walden Pond not far from Boston in Massachusetts, Thoreau used the opportunity to reflect on the underlying meanings of the emerging industrial order. His book, *Walden*, was more or less unread in its own day, but, for later generations, it became a foundational text for what has grown into environmental movements. Thoreau later contributed to the development of a new ecological science, as well, in his writings in natural history and travel diaries and notebooks. He was a transition figure in his approach to knowledge making, rejecting the cold truths of the dominant professional scientific culture to create a more explicitly personal and compassionate science of his own.

After his time at Walden Pond and the commercial failure of the book he had written about his experiences there, Thoreau gradually altered the tone of his writing. In the 1860s, he practiced a form of science that has only now, a century and a half later, begun to be properly appreciated and understood. For one thing, he became much more committed to detailed observation and systematic empirical investigation. For another, he replaced the philosophical language and literary style of

The Hybrid Imagination:
Henry David Thoreau (1817-62)

• a "romantic" scientist, author of *Walden*
• one of the founders of environmentalism
• also wrote *On the Duty of Civil Disobedience* (1849)

presentation that had characterized his earlier writings with a more scientific vocabulary and writing style. He was influenced by Darwin's theory of evolution, and he tried to bring more rigor into his own form of empathetic knowledge making. In the detailed nature journals and perhaps especially in the remarkable essay, "The Dispersion of Seeds," which was only published in the late 20th century, Thoreau had taken on what we might call a hybrid identity – the social critic and the scientist had been combined into an ecologist, even though he used the word, naturalist [Walls, L., 1995].

4.5 INDUSTRIAL SOCIETY AND ITS SOCIAL MOVEMENTS

Many of the important technological innovations of the first period of industrialization were developed in specific settings for specific purposes, but as industrialization progressed, production and the process of innovation became more systematic and complex. The horizontal and vertical integration of the capital-goods industry, the standardization of parts and the coming of the "American system of production"[Hounshell, D., 1984] starting with the production of instruments of warfare were especially important in the expansion of industrialization across the European and American continents. In what the historian Hobsbawm, E. [1975] called the "age of capital," the development of new power sources and industrial materials – especially oil and steel – were also crucial components. These innovations, along with railroad locomotives, steamships, and telegraphic com-

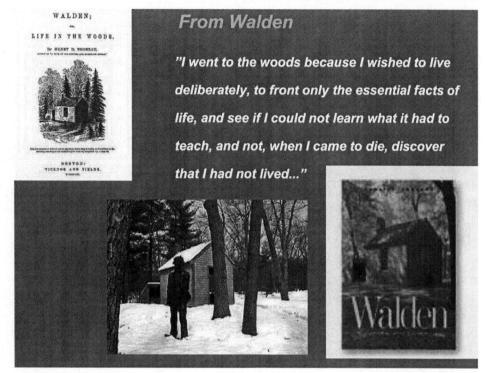

From Walden

"I went to the woods because I wished to live deliberately, to front only the essential facts of life, and see if I could not learn what it had to teach, and not, when I came to die, discover that I had not lived..."

munication, provided the material base for the consolidation of what institutional economists have termed "national systems of innovation" by which inventiveness and creativity were transformed into economically significant activities (cf. Lundvall, B., 1992). These systems involved a range of new institutions - educational, financial, legal, as well as technical and scientific – all of which were organized somewhat differently in each industrializing country.

But like the first wave of industrialization, the expansion of the industrial economy would enter a period of decline in the 1860s and 1870s, at the same time as new kinds of social and cultural agency would be mobilized in the socialist and populist movements. Like the romantic and cooperative movements, these movements emerged in protest of the injustices and iniquities of industrial society, and as they developed a cognitive praxis of their own, they carved out public spaces for new kind of knowledge-making in both science and technology.

Of all the people who contributed to the fashioning of the central discourses of the social movements that emerged in industrial society, perhaps no single person was more important than Karl Marx. Marx drew on the writings of political economists, idealist and positivist philosophers, and not least those he called "utopian socialists," such as Robert Owen in England and Charles Fourier in France, to develop a powerful intellectual synthesis that provided a narrative of science-based progress that many different types of people could share. In many ways, Marx was an intellectual

> ● ● ● | **The Second Cycle**
>
> ○ "the age of capital" (ca 1830-1880)
>
> ○ railroads, telegraph, and steel
>
> ○ technologies of socialization
>
> ○ the rise of the corporation (Carnegie, Krupp)
>
> ○ social and cultural movements:
>
> • socialism and populism
> • science fiction and arts and crafts

hybrid, combining philosophy and economics, positivism and socialism, and not least science and society.

An important source of the Marxian synthesis was the theory of natural evolution that Charles Darwin presented in his book, *On the Origin of Species*, published in 1859. Where Darwin had tried to derive from his many observations of natural phenomena fundamental laws of evolution, Marx saw himself as propounding laws of social evolution. Both were enormously influential in framing the scientific theories of the 20th century – Darwin in the natural realm and Marx in the human. Marxism became the dominant ideology of the social democratic parties and organizations that were established to represent the working classes throughout Europe. On the institutional level – especially in relation to the creation of state, or public institutions, for the regulation, administration and eventually financial support to technology and science – social democratic and, later, after the Bolshevik revolution in Russia, communist governments would play a pioneering role.

Sociology and the other social sciences were founded in the late 19th century, to a large extent in response to the writings of Marx and his followers, as part of a more general institutionalization of science and technology. The class conflict that was so central to the Marxian theories was conceptualized in more academic terms, as a transformation from *Gemeinschaft* (community) to *Gesellschaft* (society) by Ferdinand Tönnies in 1887 in his influential formulation and as a tension

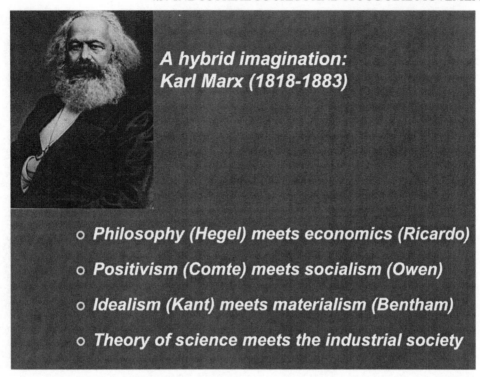

A hybrid imagination:
Karl Marx (1818-1883)

o *Philosophy (Hegel) meets economics (Ricardo)*

o *Positivism (Comte) meets socialism (Owen)*

o *Idealism (Kant) meets materialism (Bentham)*

o *Theory of science meets the industrial society*

between "organic solidarity" as opposed to "mechanical solidarity" by Émile Durkheim in 1893. The important works of Max Weber are, in many ways, also attempts to turn the Marxian ambition to develop a science of society into a legitimate and socially accepted academic practice. The social science disciplines that were established in the late 19th and early 20th centuries would develop methods and theories for explaining social change according to a disciplinary division of labor and at a certain academic distance while Marx, as a movement intellectual had sought to develop his science in explicitly political and value-laden terms.

For Weber, industrialization had brought about a general process of rationalization, both in the economy, the state and what started to be called civil society. As such, while science had grown into what he later would characterize as a vocation, it had also lost something of its enlightening or cultivating function. His response was to carve out a specialized identity as a social scientist and create what Pierre Bourdieu later called the habitus of the academic discipline, while many others took their scientific and technological competence directly into the commercial marketplace. As in earlier periods in history, the hybrid imagination that had been fostered in the social and cultural movements of the times tended to split apart into the contending cultures of science and engineering.

4.6   WILLIAM MORRIS AND ARTS AND CRAFTS

One of the more influential "movement intellectuals" in the second cycle of creative reconstruction was William Morris and what might be termed the sub-movement that came to be called the Arts and Crafts Movement. William Morris started out as a romantic poet and, by his own account, a dreamer, but in the course of his life, he came to mix artistry and practical skills to create the identity of the designer, which is how he identified himself. And it is as one of the first designers that he continues to serve as a multifaceted source of inspiration for the cultural appropriation of technology and science in the 21st century.

Morris's decidedly ambivalent attitude to technology and science was formed by experience, and in particular the experience of making beautiful things for a living. Since the 1850s, when he and a few of his friends from Oxford had formed the "Firm" - which later became Morris and Company - he had worked in the interior decoration business, and in the process, he had mastered a wide range of practical skills and learned to use a great variety of tools and sometimes even machines. While his wallpapers, tapestries, carpets and furniture had become fashionable consumer items in Victorian England, he had grown increasingly frustrated by the fact that they were too expensive for the working people to whose revolutionary cause he had become committed. Starting in the 1860s, he had begun to give public lectures on the relations of art and society, and he had joined into

protests over the destruction of old buildings in London. In the 1870s, he became an active member of the Socialist League, one of the precursors to the Labour Party [MacCarthy, F., 1994].

Toward the end of his life, in 1890, Morris wrote a utopian vision of the future, entitled *News from Nowhere*. It was not meant to be great literature but rather to contribute to the struggle for socialism. It was first published in installments in *Commonweal*, the magazine he had helped start a few years earlier, and it would later be bound in leather with the full range of engravings and typographical innovations that Morris developed at his Kelmscott Press. *News from Nowhere* was written both to counter the traditionalism and "anti-scientific" attitudes that were widespread among many of his fellow British artists and writers, as well as the futurism and technological optimism that was endemic to most of his fellow socialists. More specifically, it was meant as a response to the attitudes to technology and industrial society that had been articulated by Samuel Butler in his book, *Erewhon* (1872), on the one hand, and by the North American Edward Bellamy in *Looking Backward* (1888), on the other. Where Butler, in accordance with many romantic artists, had imagined the decline of technological civilization and, in his novel, had placed all machines into a museum, so they could no longer cause any trouble, Bellamy, the American populist writer, had envisioned the enormous expansion of industrial machinery into ever more areas of life, with production in the hands of an industrial army in which all were expected to serve. For Morris, both visions were fundamentally flawed, the one in its romantic pessimism and the other in its populist optimism.

A central theme in the many lectures that Morris gave on the relations between art and socialism was the need to make proper use of machinery. In "Useful Work versus Useless Toil," which he delivered on several occasions in 1884 and later published as a pamphlet, he put it this way:

> Our epoch has invented machines which would have appeared wild dreams to the men of past ages, and of those machines we have as yet *made no use.* They are called "labour-saving" machines – a commonly used phrase which implies what we expect of them; but we do not get what we expect. What they really do is to reduce the skilled labourer to the ranks of the unskilled [Morris, W., 1993, p. 304].

Morris was particularly critical of the degeneration of craftsmanship in what he liked to call the "lesser arts" due to the greed and indifference of the industrial capitalists. In a typical formulation from a lecture given for the Society for the Protection of Ancient Buildings in 1882 on 'The Lesser Arts of Life," he castigated the "rejecters of the arts who are corrupters of civilization:"

> You understand that our ground is that not only is it possible to make the matters needful to our daily life works of art, but that there is something wrong in the civilization that does not do this: if our houses, our clothes, our household furniture & utensils are not works of art, they are either wretched makeshifts, or, what is worse, degrading shams of better things [Morris, W., 1882].

In the arguments of Morris, we find the seeds for contemporary criticisms on the ever increasing number of goods and products of poor functional and aesthetic quality that pervades the globalized commercial marketplace. For Morris, it was a regeneration of the arts in the name of socialism that would provide a constructive way for dealing with the problems of industrial society. As a precursor of contemporary environmentalism, he also urged that the arts should be pursued with what we called a love of nature, which he himself exemplified in the many designs for wallpapers and tiles that he made with animals and plants conspicuously visible:

> love of nature in all its forms must be the ruling spirit of such works of art as we are considering; the brain that guides the hand must be healthy & hopeful, must be keenly alive to surroundings of our own days, & must be only so much affected by the art of past times as is natural for one who practices an art which is alive, growing, and looking toward the future [Morris, W., 1882].

As we shall see in the following chapter, Morris's ideas about design and the practical, or lesser arts would inform many of the "modernist" projects of the 20th century, both among artists and architects, as well as among planners and politicians, both in Europe and North America. His portrait hung in Hull House in Chicago, where Jane Addams sought to alleviate the distress of the urban poor, and many of the social reformers of the 1930s, the makers of modernity and the welfare state, returned to his lectures and *News from Nowhere*.

Already in his own lifetime, Morris was a central figure in what came to be called the Arts and Crafts Movement [Cumming, E. and Kaplan, W., 2002]. In Germany, Austria, and Scandinavia, as well as in Britain and North America, the various branches of the arts and crafts movement that built on the works and ideas of Morris brought about a kind of revolution in the way in which products are made. Above all else, it was the emphasis on quality – in production as in life - rather than the reduction of quality to a quantitative, mechanical logic that remains the "legacy" of William Morris in our day.

Like all movements, the ones that were dedicated to the promulgation of arts and crafts split into a wide range of disparate groups as they became institutionalized, and the direct influence of Morris on the further development of technology and science was highly differentiated. Yet the idea of bringing artistic energy into industrial activities, mixing business with pleasure, so to speak, lives on in many an architectural firm, and not least in many a designer's workplace. Similarly, Morris's organicism, the artistic representation of natural processes that made his wallpapers and tapestries so appealing in late Victorian England has continued to exert a significant influence on designers and architects, from Frank Lloyd Wright to Frank Gehry.

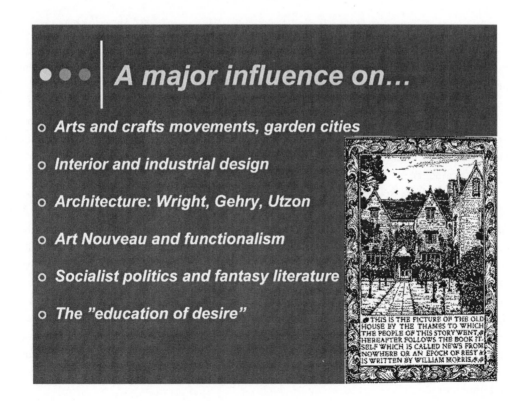

A major influence on...

o **Arts and crafts movements, garden cities**

o **Interior and industrial design**

o **Architecture: Wright, Gehry, Utzon**

o **Art Nouveau and functionalism**

o **Socialist politics and fantasy literature**

o **The "education of desire"**

For other artists and architects, especially those captivated by technology and its rationality, the example of William Morris has taken on an anachronistic, backward-looking meaning. In rejecting the organic and accepting the instrumental imperatives of contemporary life, many a modernist has come to see the artisan tradition that Morris epitomized as a barrier to innovation. "Functionality" became the artistic catchword, but in becoming established, it has often been connected to a form of hubris, a tendency to building or designing technically daring but aesthetically ugly structures and products. Under the spell of hubris, many modern artists and architects have bestowed upon us the fantastic constructs of their imaginations while they have strayed ever further from their Morrisian roots.

4.7   THE POPULIST SENSIBILITY

The critical social movements that emerged in the United States in the late 19th century were less directly associated with socialism and Marxism than was the case in most European countries. Instead, particularly in the rural areas of the country where small farmers and businessmen were struggling to survive, a populist movement – often with a religious flavor - developed, to oppose the Eastern trusts and railway tycoons and propose alternative approaches to technology and science [Kazin, M., 1995].

The populist influence on technology would be significant – both in terms of fostering new uses for technical innovations, as well as in opening educational and scientific opportunities for the rural population. In the United States, both the land-grant, state universities and the smaller private colleges were influenced by the ambition to make knowledge making and technological development popular, accessible to the entire population. The strong social orientation of many scientific and engineering projects at the turn of the century – from the geological surveys to the urban sociology that developed in Chicago under the auspices of Jane Addams – owed much to the populist movement, as it exercised influence over the Democratic Party, as well as many local and state governments.

Several of the key technological innovations that would be so important in the next phase of economic growth and expansion were also connected to the populist sensibility of the American Midwest. Both Henry Ford and the Wright brothers can be said to have developed their self-confidence and their self-taught approach to technological experimentation as part of a populist cognitive praxis. In any case, their belief that they could do more effectively what the professional scientists and engineers "back East" could not do, as well as their ability to combine social and economic understanding with technical skills and scientific knowledge certainly owed something to the political culture in which they lived. As we will discuss in the next chapter, the manifestations of a hybrid imagination were quite different in the two cases, with the Wright brothers trying to hold onto theirs well after their momentous achievement in conquering the air, while Henry Ford would come to be the very epitome of hubris after making his radical innovations in automobile design and production.

The Wright brothers and their flying machine

Orville and Wilbur Wright made bicycles in a shop in Dayton, Ohio, where they had grown up, and they were largely self-educated in matters of technology and science. They were, however, dedicated experimenters, and showed themselves to be extremely adept in the arts of problem-solving. But perhaps most significantly, they were idealist Christians who were interested in solving the problems involved in making an airplane, not for reasons of personal gain but out of a sense of what might be called social service.

As Tobin, J. [2003] brings out in his recent book, *First to Fly* the success of the Wright brothers over their many rivals was due, in large measure, to their having a hybrid imagination. While Samuel Langley, Alexander Graham Bell and other well-established scientists and engineers took on the challenge of making an air machine with an interest primarily in achieving fame and fortune very much in mind, for Wilbur Wright and his younger brother Orville, their work was more a fulfillment of a childhood dream and a way to have some fun and meaningful relief from their bicycle business. As Wilbur had written to his father in 1894 to ask for a loan so that he could go to college:

> I do not think I am specially fitted for success in any commercial pursuit even if I had the proper personal and business influences to assist me. I might make a living but I doubt whether I would ever do much more than this. Intellectual effort is a pleasure to me and… I have always thought I would like to be a teacher. Although there is no hope

of attaining such financial success as might be attained in some of the other professions or in commercial pursuits, yet it is an honorable pursuit, the pay is sufficient to live comfortably and happily… (quoted in Tobin, J., 2003, p. 46).

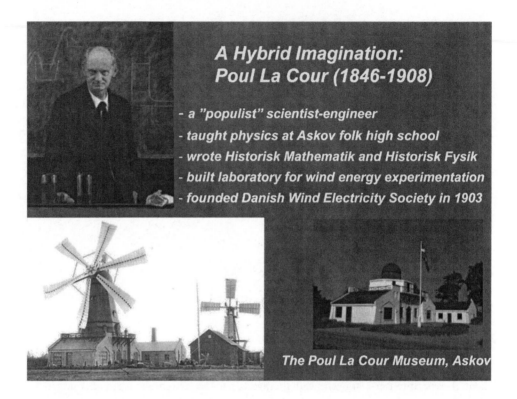

A Hybrid Imagination:
Poul La Cour (1846-1908)

- *a "populist" scientist-engineer*
- *taught physics at Askov folk high school*
- *wrote Historisk Mathematik and Historisk Fysik*
- *built laboratory for wind energy experimentation*
- *founded Danish Wind Electricity Society in 1903*

The Poul La Cour Museum, Askov

Although his father was willing to give him the loan, Wilbur Wright never went to college. Perhaps, writes Tobin, he thought he should pay his own way. In any case, he did pursue his intellectual effort and spent most of his life as a public educator, teaching others how to make airplanes. The way in which he and his brother approached the problem was to observe closely and precisely how birds kept their balance in flight and then try to model their ever more sophisticated aircraft designs on what they had seen. They combined dedicated reading in the scientific and technical literature with a series of experimental tests, rather than trying to apply a scientific theory or construction technique that had been used in other areas, as did their rivals. Their hybrid imagination, and not least their ability to complement each other and work together, succeeded where the hubris of their rivals failed.

Different forms of populism and socialism emerged in other parts of the world, as well, in the late 19th century with different relations to science and technology. In Denmark, the farmers' movement, as it has been called, was an important site for technical development, both in relation to agricultural machinery, but also in relation to wind energy and food-processing technology. A system of state-supported technical consultancy was set up, along with the spreading of people's high schools in the countryside, and it was at one such school that Poul La Cour both developed what might be termed socially-informed ways to teach mathematics and physics, by placing them in historical perspective, but also by pioneering in the use of electricity to produce wind energy – which would plant important seeds in the Danish society that would come to be so successfully harvested in the 1970s [Jamison, A., 1982, p. 271].

CHAPTER 5

Science, Technology and Modernization

We would not destroy the rigorous method of science or the resourceful technology of the engineer. We would merely limit their application to intelligible and humane purposes. Nor would we remove altogether the mechanical world-picture, with its austere symbolism; we would rather expand it and supplement it with a vision of life which drew upon other needs of the personality than the crude will-to-power.

Lewis Mumford, "Toward Civilization" (1930), quoted in Miller, D. [1989]

5.1   THE THIRD CYCLE

At the end of the 19th century and in the first years of the 20th – the so-called "turn of the century" – a number of new technological artifacts entered the world. And as they did, the world would change in some rather fundamental ways. The telephone (or electrical speech machine, as Alexander Graham Bell called his invention), automobiles (or horseless carriages as they were referred to at the time), airplanes (then known as flying machines), and perhaps most significantly of all, phonograph machines (remember them?) were intrinsically different from those of the previous "waves" of industrialization in the 19th century. They were not simply improvements, however sophisticated, on things that people had used for centuries. They were rather figments of a collective hybrid imagination, by which scientific facts were literally implanted into technological artifacts.

These were technologies that were intimately connected with science, and they were, indeed, unimaginable without scientific knowledge or at least large doses of science-based thinking and experimentation. The automobile was based on a new kind of motor, the internal combustion engine that had been developed at technological universities in Germany after some of the basic laws of thermodynamics had been formulated; to finally make machines that could fly through the air after many centuries of fruitless trying required scientific knowledge in such fields as aerodynamics, atmospheric science and, even, as it turned out, animal physiology in relation to how birds do it;

and the telephone and the phonograph drew on scientific work in both mathematics, acoustics and electromagnetic theory, to name only a few of the relevant fields that supplied facts to the makers of the technological artifacts in question.

In organizational and institutional terms, it had been the systematic cultivation and professionalization of scientific research and the formation of new fields of engineering science that had taken place in the second half of the 19th century that had made the inventions possible. In the course of industrialization, new contexts, or sites for making science and technology had been created in the form of industrial research laboratories and university engineering departments, and it was in such places that a good deal of hybrid imagining was carried out, which contributed to the technological breakthroughs. Even though the successful innovators, such as Alexander Graham Bell, Henry Ford, Thomas Edison and the Wright brothers, who were able to bring the inventions into the society and the commercial marketplace, were not professional scientists or engineers, they had definitely made use of science in their inventing activity.

● ● ● | *The Third Cycle*

○ *"the age of empire" (ca 1880-1930)*

○ *Electricity, automobiles, and airplanes*

○ *Technologies of modernization*

○ *Technology becomes incorporated (GM, GE, AT&T, etc)*

○ *Social and cultural movements:*

· *modernism and anticolonialism*

· *technological civilization and mass culture*

At the same time as these new technological artifacts were entering the world, there occurred a fundamental reformulation of the basic "facts" of science themselves in the four papers that the 26-year old Albert Einstein published in his "miraculous year" of 1905, most significantly, in the long run, the special theory of relativity. A few years later the young Dane, Niels Bohr, presented his atomic model with the now familiar electrons and neutrons acting on each other in "indeterminate" ways as the basic component parts of material reality. In the coming years, Bohr and Einstein would be the leading figures in rewriting the "book of nature," creatively reconstructing the underlying world-view assumptions that had dominated science and engineering since the 17th century. Einstein's achievements, in the words of his most recent biographer, "reflected the disruption of societal certainties and moral absolutes in the modernist atmosphere of the early 20th century [Isaacson, W., 2009, p. 3]." Indeed, in both his science as well as in his lifelong quest for peace and social justice, Einstein was very much involved in what we will be characterizing in this chapter as a modernist movement.

A Hybrid Imagination:
Albert Einstein (1878-1955)

"Imagination is more important than knowledge"

Einstein's life and works can be seen as the outpourings of a particularly fruitful hybrid imagination. As Galison, P. [2003] has recently argued, Einstein's radical reconstruction of our conceptions of time and space combined a highly-developed imagination with a more mundane, down-to-earth practical understanding of the working of clocks and other artifacts, which he had gained in going through applications for patents at his place of work, the patent office in Bern. Later, during and after the first world war, Einstein would seek to combine his scientific and technical competence with a quest for social justice, in attempting to tame the hubris of Nazism and, for that matter, the quantum theorists that he never accepted, and his "hybrid imagination" came to provide a source of inspiration for scientists and engineers – and the rest of us - ever since.

In addition to the technological and scientific innovations, there were also a number of social, or organizational innovations that characterized the third wave of industrialization, and which encouraged a tendency to hubris in the decades to follow on the part of many a scientist and engineer. On the one hand, there was the incorporation of science and technology in new kinds of research and development companies that had started with Thomas Edison's famous invention factory in Menlo Park, New Jersey in the late 19th century, and which became institutionalized in the early 20th century in such firms as General Electric, Dupont, and General Motors. On the other hand, there were the methods of mass production and the so-called assembly line, pioneered by Henry Ford, and which would become extremely important in the subsequent period of economic growth and expansion

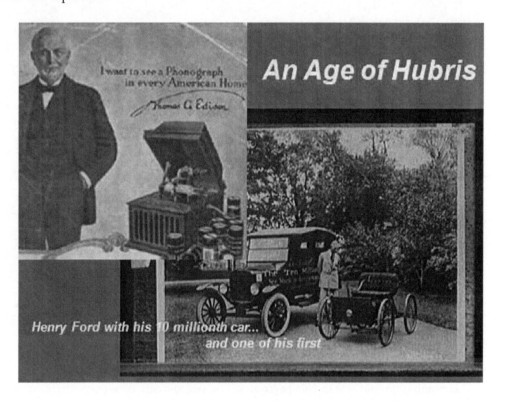

Ford's conveyor belts provided a centralized approach to the production process and, not least, to the distribution and marketing of products. Frederick Winslow Taylor's time and motion studies and more general program of scientific management provided a systematic approach to the use of so-called human resources, and thereby the possibility of a much more ambitious rationalization of production and "human engineering" than had previously been attempted. The new century thus encouraged a widespread spirit of hubris in which technology and science were seen as the harbingers of a new kind of modern civilization that replaced non-human nature with man-made things.

The "American system of production" as it came to be called was not confined to Bell's, Ford's, Taylor's, Edison's and the Wright brothers' innovations [Hounshell, D., 1984]. It also included the systematic utilization of interchangeable parts across the industrial landscape, as well as the spread of new manufacturing and building materials such as steel and other alloys, and new sources of energy, such as oil. Mass production also engendered mass consumption, and in the first half of the 20th century, new forms of advertising and marketing became important features of contemporary life and contemporary education. Industrial society had grown into the mass society of a modern technological civilization.

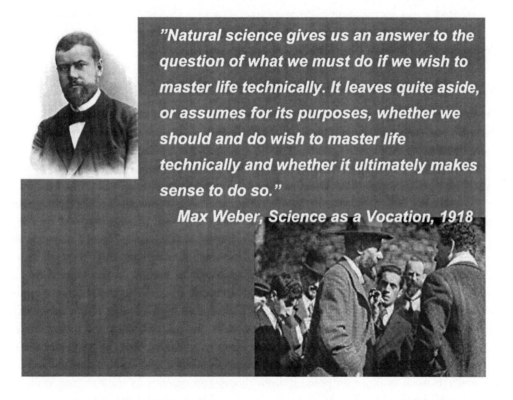

"Natural science gives us an answer to the question of what we must do if we wish to master life technically. It leaves quite aside, or assumes for its purposes, whether we should and do wish to master life technically and whether it ultimately makes sense to do so."
Max Weber, Science as a Vocation, 1918

The American system would spread around the world during the first decades of the 20th century as the exemplary model of modernity, even to the new Soviet Union. But while the new production technologies and organizational methods and technological artifacts were widely admired and copied, the cultural ideals associated with the new technologies were not so readily accepted. Indeed, there was concern that the American-dominated technological civilization would involve a general decline in cultural values. As Max Weber put it in 1918, in a lecture for students at the University of Berlin, "Science as a Vocation:"

The young American has no respect for anything or anybody, for tradition or for public office - unless it is for the personal achievement of individual men... The American's conception of the teacher who faces him is: he sells me his knowledge and his methods for my father's money, just as the greengrocer sells my mother cabbage... no young American would think of having his teacher sell him a *Weltanschauung* or a code of conduct [Weber, M., 1958, pp. 149–150].

While Americans generally welcomed modern civilization, many Germans and other Europeans bemoaned the loss of what they termed *Kultur* and the values associated with the past. In the 1920s and 1930s, this sense of cultural decline would lead to what the historian Herf, J. [1984] has called "reactionary modernism." Like the modernist movement in general, reactionary modernism represented a way to combine cultural awareness with scientific and technological development. In this chapter, we will try to show how the modernist movement contributed to the subsequent development, or creative reconstruction of science and technology.

5.2 THE MOVEMENTS OF MODERNITY

We have seen in the previous chapter how the social and cultural movements of the 19th century – the romantic and cooperative movements of the early part of the century and the socialist and populist movements of the latter part – provided public spaces for the mixing of critical ideas and constructive practices in new combinations. In our terms, they served as sites, or cultural contexts for fostering hybrid imaginations that contributed to the creative reconstruction of science and technology.

In the first half of the century, significant scientific discoveries, such as electromagnetic current, and key technological inventions, such as the telegraph, were directly inspired by the romantic movement. People like Hans Christian Ørsted in Denmark and Henry David Thoreau in the United States sought to bring the romantic "sense of wonder" directly into their research as an explicit alternative to the mechanical or instrumental approach to nature. In the same time period, the more politically-oriented cooperative movement served as a breeding ground for new approaches to science and engineering education, such as the Danish folk high schools, which were created in the countryside as a way to help keep rural values alive in the industrializing and urbanizing society.

In the second half of the century, new kinds of scientific facts, in particular the extremely influential works of Karl Marx in social and economic science, and new ways of making technological artifacts, such as those proposed in both theory and practice by the designer William Morris, were directly inspired by the socialist movement. In Denmark and the United States, the populist movement was an important source of inspiration for the pioneering experimentation with wind power-generated electricity by Poul la Cour, a physics teacher at Askov folk high school, and the invention of the airplane by the Wright brothers. As in the first cycle, the socialist and populist movements also served to inspire new approaches to science and engineering education, such as the regional technological universities and the state land-grant universities in the United States where engineers could bring science into the surrounding society and the schools for designers and artisans that grew out of the arts and crafts movement.

In the subsequent periods of consolidation and institutionalization, the integrative "cognitive praxis" of the social and cultural movements was transformed into contending cultures of science and engineering. While a few scientists and engineers tried to retain their hybrid identities in the next "wave" of industrialization, many more were attracted by the lure of fame and fortune and tended to become more commercial, or entrepreneurial in their attitudes and behavior. Others were pulled back by forces of habit(us) into earlier and more traditional ways of making knowledge as part of a more general reaction, or "counter-movement" to industrial growth and expansion. In the name of national or ethnic or cultural "purity," reactionary counter-movements mobilized the traditions of anti-hybridity that had earlier resisted the hybrid imaginations of the Renaissance and the Enlightenment. It was also in these counter-movements that ideas of "pure science" began to be articulated in the 19th century that would be mobilized again in the counter-movements of the 20th and 21st centuries, among the anti-communists in the 1950s, and the climate skeptics in our day.

From such a perspective, the 1910s and 1920s also marked a period in which social and cultural movements had a decisive influence on the creative reconstruction of science and technology. A wide-ranging and highly diverse modernist movement made fundamental contributions to shaping what in the 1930s and 1940s grew into the various societal programs of modernization. For all of their differences, not least ideologically, modernists sought to give human meaning to what was widely seen, especially after the destruction of the first world war as an inhuman and barbaric technological civilization. Modernism in its many guises – from liberally progressive to explicitly reactionary – was particularly important in bringing artists and other humanists with their cultural sensibility into contact with scientists and engineers.

In the same time period, anti-colonialism provided a seedbed for a more culturally appropriate science and technology to develop in those parts of the world that had been under the domination of European imperialism. As we shall see, the Indian struggle for independence involved at its very core movement intellectuals with hybrid imaginations, in particular, Mohandas Gandhi and Rabindranath Tagore, who sought to combine the modern science and technology of Western civilization with an appreciation for indigenous, non-Western traditions and approaches to knowledge-making. In India, modernity took on a different meaning than it did in the industrialized countries, largely because of the efforts of Tagore and Gandhi and the cognitive praxis of the anti-colonial movement. After independence, both Tagore and Gandhi have continued to serve as role models for hybrids throughout the world, perhaps especially among those who left the sub-continent after the disastrous separation of India and Pakistan in 1949. The Nobel prize-winning economist and philosopher of justice, Amartya Sen, has recently written about how Tagore was a particularly important source of inspiration for him [Sen, A., 2005].

As in the 19th century, the ideas and practices of modernism and anti-colonialism were transformed as they spread or diffused into the broader society. In the 1930s and 1940s, there was a strong tendency to hubris in the various processes of modernization, and there was, as well a resurgence of anti-hybridity in the reactionary movements that took over state power in Germany, Spain, Italy and Japan. As such, the hybrid imaginations that had been fostered in the movements of the 1910s and 1920s were forced to contend with the incorporation, or "mainstreaming" pressures of a dominant commercial culture, on the one hand, and the pull of a reactionary and traditionalist residual culture, on the other.

5.3 WHAT WAS MODERNISM?

In the words of Peter Gay, who has written one of its most recent histories, "modernism is far easier to exemplify than define," but he then goes on to say, by way of definition that:

> For all their palpable difference, modernists of all stripes shared two defining attributes…: first, the lure of heresy that impelled their actions as they confronted conventional sen-sibilities; and, second, a commitment to a principles self-scrutiny. Other possible criteria of classification, no matter how promising, all failed: political ideologies, though invit-ing, cannot serve to define modernism, since it is compatible with virtually every creed, including conservatism, indeed fascism, and with virtually every dogma from atheism to Catholicism [Gay, P., 2009, pp. 3–4].

The various wings of the modernist movement were not always compatible with one another, to say the least. As we shall see, some were avowedly futuristic in their enthusiasm for modern technology, while others were passionately opposed to what grew in the 1920s into a new "techno-cratic" faith. There were progressives, pragmatists, conservationists, regionalists, and a good number of outright mystics within the modernist movement.

While differing in many ways, the various wings of the modernist movement shared an attitude of what we have characterized as critical engagement. They were all aiming, in one way or another, to steer or move scientific and technological development into what were considered more culturally appropriate directions. They were trying to humanize the machine, to bring values back into what was increasingly perceived as a overly mechanical and instrumental society. In what Gay calls their heresy, they shared an ambition to reject all that was old and create new things in every sphere of life, and in their self-scrutiny, they had a common interest in reflecting on the meaning of what they were doing. As such, there was, among them, many a hybrid imagination in action. They mixed the arts and, more generally, a humanist sensibility into the making of science and technology. And as the movement grew into the institutions of modernity, there would be both market-driven and community-oriented versions that would together do battle with the forces of anti-modernism. In the words of Raymond Williams:

> Although Modernism can be clearly identified as a distinctive movement, in its delib-erate distance from and challenge to more traditional forms of art and thought, it is also strongly characterized by its internal diversity of methods and emphases: a restless and often directly competitive sequence of innovations and experiments, always more immediately recognized by what they are breaking from than by what, in any simple way, they are breaking towards. Even the range of basic cultural positions within Modernism stretches from an eager embrace of modernity, either in its new technical and mechanical

forms or in the equally significant attachments to ideas of social and political revolution, to conscious options for past or exotic cultures as sources or at least as fragments against the modern world, from the Futurist affirmation of the city to [the poet T.S.] Eliot's pessimistic recoil [Williams, R., 1989, p. 43].

In the 1910s and 1920s, modernism was joined by the more explicitly political movements of anti-colonialism in Latin America, Africa, and Asia, which brought non-Western voices and perspectives into the mix. And like all movements, modernism and anti-colonialism were temporary and relatively brief historical moments "in between" waves of growth and expansion. As the movements became institutionalized in the 1930s onward, amid preparations for what would become a second world war, there was, among both modernists and anti-colonialists, a strong tendency to hubris that entered into their lives and actions, most notably in the making of the atomic bomb. As we look back at the 20th century, however, there can be little doubt that the various interpretations of the modern condition that were articulated and practiced within the modernist and anti-colonial movements were formative for the ways in which humanity came to interact with science and technology.

On the front-page of *Le Figaro* on February 20, 1909, there appeared a manifesto written by the Italian writer Filippo Tommaso Marinetti that, at least, retrospectively, has come to mark the beginnings of the modernist movement. Just the fact that a cultural document appeared on the front page of one of the leading newspapers in Europe indicates the importance of the provocative manifesto, which, in its famous final paragraph, provides a kind of secular hymn to a science-based technological civilization:

> We shall sing… the nocturnal vibration of the arsenals and the yards under their violent electrical moons; the gluttonous railway stations swallowing smoky serpents; the factories hung from the clouds by the ribbons of their smoke; the bridges leaping like athletes huled over the diabolical cutlery of sunny rivers; the adventurous steamers that sniff the horizon; the broad-chested locomotives, prancing on the rails like great steel horses curbed by long pipes, and the gliding flight of airplanes whose propellers snap like a flag in the wind, like the applause of an enthusiastic crowd [Marinetti, F., 1960].

Marinetti had not been to the United States, but in his colorful, metaphorical language there is a strong resemblance to what Nye, D. [1994] has termed the American "technological sublime" - the widespread feeling of awe and worship, sometimes mixed with terror - that developed in the latter part of the 19th century and was still going strong at the beginning of the 20th as the American technological civilization came into its own. There can be little doubt that Marinetti was acquainted with the American poet Walt Whitman's *Leaves of Grass* because both the form and content of the quoted paragraph from the manifesto is very close to the way in which Whitman "sings the body electric" in one of the poems in his famous work.

Futurism represented the onset of a period of revolt and revolution against the past, more specifically, the old European social and political order on behalf of an American-inspired, science and technology based future. It was in large measure an artistic revolt because it was in the arts that the modernists would look for the tools and methods for appropriating science and technology for the liberation and emancipation of mankind. Art was to be joined with technology to transform humanity into something truly new and different; the artist and the engineer would combine forces and competence in a new kind of hybrid identity as the change agents of modernity. The artist, breaking down the barriers of tradition and the engineer, creatively destroying the artifacts of the past would become one.

Modernism, in the aftermath of the devastating wars of the 20th century, has come to be seen with critical eyes. The emphasis on war as "ultimate hygiene" of culture and race, which came with the reactionary modernism of the Nazis, the masculine renunciation of alleged female qualities as docility, fragility, and empathy as corruptive forces, and the faith in technological progress have all been seen as hubristic, especially as scientific and technological development have become increasingly part of military-industrial complexes, both mental and material. In its own time, the futurist fascination with the technological sublime was an important influence in the cultural appropriation of science and technology that took place in the first few decades of the 20th century.

A new kind of art emerges

- *appropriating technology and science*
- *experimenting with light and color*
- *representing "impressions" of reality*
- *using machines as metaphors*
- *trying to visualize motion and abstraction*
- *making ordinary things beautiful*

Experimentation with new forms of artistic technique and expression had been widespread since the late 19th century. Monet and Van Gogh, Gaugin and Cezanne – to name only some of the more influential impressionists and post-impressionists – had sought to create a new kind of art that challenged the established routines and the traditional ways of depicting reality. These artists no longer tried to paint reality as it really was – photographs could do that so much more effectively. Rather, they tried to provide through their art some of the "structures of feeling" that were appropriate for a technological civilization. Their task was to conjure up meaningful impressions of the world, rather than accurate pictures of reality.

The modernist movement, in its efforts to bring a kind of cultural meaning into science and technology, built on these precursors, and it served to give art a new role in the modern world, at one and the same time popular and sophisticated. They made space and time, nature and society, object and subject, and, not least, the conscious and the unconscious into central topics of artistic exploration. It was an art that could appeal to both highbrow and lowbrow, both the masters and the masses, and there was no question about its modernity. Mixing genres and techniques, modern artists and architects and eventually designers and film-makers mobilized the traditions of the Renaissance, when art and engineering were combined in hybrid imaginations.

It was in the modernist movement that Paul Klee, Joan Miro, Marc Chagall, Pablo Picasso and so many others would develop a totally new kind of art, by seeking to appropriate with their artistic skills the new scientific facts and technological artifacts of their times. Einstein's cosmology of space and time had a strong influence on the move to abstraction that was taken by the pioneering artists

Rene Magritte: mixing realism and illusion

of the modernist movement, while another school of modern art - the surrealists: Salvador Dali and Rene Magritte, among others - was inspired both by physical theory as well as psychoanalytical theory to depict a dream-like universe of the mind far removed from the realism of the artists of the pre-modern era.

5.4 A HYBRID IMAGINATION IN ACTION: THE BAUHAUS

In Germany, a kind of organic engineering aesthetic was manifested in the *Bauhaus* of the Weimar years, as well as in expressionist art, as artists and engineers sought to bring their talents to bear on the technological civilization. The German architect Walter Gropius had been a sergeant major in the imperial army during the first world war and had lived through the atrocities of the trenches on the Normandy front. The experience had a crucial impact on his thoughts as he formulated his own manifesto in 1918. The metaphor that he used to describe what would be his school for artists and designers was the "Cathedral of Socialism," mixing a political concept with a religious image. In looking back to the cathedrals of the middle ages, he was inspired by William Morris and the arts and crafts movement, but in his aesthetics, he was also inspired by the rationalism and functionalism of 20th century science and technology.

Because of their wartime experience and the destructive uses of science and technology that they witnessed first hand, Gropius and his colleagues were far more critical than the futurists had been about how science and technology were being developed and used in society. For Gropius, it was the destructive elements in the modern technological civilization that needed to be tamed by the artist employed in the design and construction of artifacts and art-works.

Gropius contended that it was interaction and interdisciplinary collaboration that would lead to a more effective and socially beneficial technological development and artistic expression: "Today they (artists) are tied up in a self-sufficient position and they can only be freed from that by conscious collaboration with all workers," he wrote in his manifesto. And he argued that there was no real difference between the artist and the craftsman; actually, the artist is a craftsman who is simply able to make things that combine functional utility with aesthetic beauty: "There is no fundamental difference between the artist and the craftsman. The artist is an upgrading of the craftsman," he wrote [Gropius, W., 1918]. Gropius used the German term *Steigerung*, which means there is some kind of inner force in the artist that takes craftsmanship to a higher level, or realm, be it ethical, spiritual, and/or aesthetic.

The early Bauhaus school that was established in the new German capital city of Weimar in 1919 was characterized by a diverse and expressive attitude toward craftsmanship and art. The pedagogy of the school was closely related to the skills and competencies of the teachers who were attracted to the program; Johannes Itten, Wassily Kandinsky, Oscar Kokoschka and Paul Klee had different ideas and artistic ideals, but what they all strived for was that functionality and rationality should be coupled with aesthetics and metaphysics in an interactive process of creative learning.

The collective ethos of the Weimar Bauhaus did not last long, however [Gay, P., 2009, Whitford, F., 1984]. Gropius increasingly emphasized the importance of bringing into the program

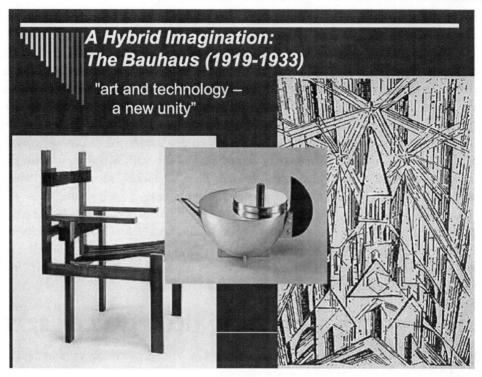

A Hybrid Imagination:
The Bauhaus (1919-1933)

"art and technology –
a new unity"

the technical, physical and functional reality of the social world, while Itten became increasingly distant in his Buddhist-inspired approach, and Klee more or less disappeared into the professional art world, and left the school. Both were replaced by teachers who had a more technical approach to art and design. In the programs, designs, architectures and pictures that resulted from the research and teaching in the school, we see a clear trajectory from a holistic and all-embracing attitude in the immediate years after the first world war to the establishment of the Dessau Bauhaus (1926) under the leadership of Hannes Meyer and Mies van der Rohe, which resulted in a much more mechanical or instrumental approach.

The intent of the school throughout its short life (it was closed down when the Nazis came to power) was to focus the combined attention of art and technology on the needs of ordinary people and how their requirements and needs could be met in a meaningful way, both in terms of function, ethics and aesthetics. The first Bauhaus school also paid a great deal of attention to spiritual values, which can be seen in the educational programs that were led by Kandinsky and Itten. Kandinsky had published a book in 1912, *On the Spiritual in Art*, which provided the point of departure for his lectures at the school, and Itten performed zen buddhist exercises in his classes as an introduction to using color, composition and light in painting. The school in Weimar was characterized by a mixture of materialism and functionalism together with the spiritual values of expressionism, and what would come to be called existentialism, that made the school's program appeal to searching and open-minded students from all over Europe.

Due to political pressure from the rapidly growing Nazi party in Weimar and the crisis of the Weimar Republic, the school moved to Dessau in 1926, and Gropius designed the new school buildings from the bottom up, making it one of the iconic structures of modern architecture on a global scale. The notions of equality, liberty and transparence are clearly present in the architectural outline and content both for the school buildings themselves, as well as for the housing for students and professors. But something happened as the school was transferred to Dessau because suddenly Kandinsky, Itten and Kokoschka were no longer part of the staff, which meant that the spiritual and expressionistic approach faded from the curriculum. The school focused in the new setting on functionalism and the practice of architecture and design, although there was still an overall socialist, or social-democratic orientation, especially in the work of Hannes Meyer. In 1926, the school entered into a close collaboration with the German airplane manufacturer, Junker, and this collaboration had certainly a role in the technical and analytical turn in the school's curriculum. The school was closed down in 1933 by the Nazi government and many of the teachers emigrated to the United States where they continued their work at the design school in Chicago.

5.5   HUMANIZING THE TECHNOLOGICAL CIVILIZATION

Although there were many attempts in the modernist movement to develop a cultural understanding of science and technology, perhaps no one person was as influential as the American "public intellectual," Lewis Mumford. Mumford was born in New York City in 1895, in what American historians have come to call the "progressive" era, a time of social reform with many a young intellectual-to-be following in the footsteps of the philosopher John Dewey and the social worker Jane Addams, to try to do something about the myriad deficiencies that were to be found in the rapidly growing and industrializing American society. Particularly serious were the living and working conditions in such places as Cleveland, Detroit, and Chicago, as waves of immigrants had come from Europe in search of a better life in the factories of the new industrial cities.

Already in his teens, Mumford started writing for the progressive magazine, *Dial* (where the progressive intellectuals, Thorstein Veblen, Randolph Bourne, and John Dewey were among the main contributors), and in the 1920s, he would take active part in envisioning another, more humanly appropriate kind of American civilization, serving as secretary of the Regional Planning Association of America, and beginning his long series of books dealing with the interconnections between technology, cities, art and literature [Hughes, T.P. and Hughes, A.C., 1990]. Mumford would seek to tame the "hubris" of the American system of production with its assembly lines and scientific management by means of historical analysis.

Lewis Mumford accepted neither the technocratic position, which was particularly strong in the United States, nor the disenchantment with technological civilization that was typical of the humanists and had been made popular by the drama critic, Joseph Wood Krutch, in his book *The Modern Temper*, from 1929. Like the poet T.S. Eliot and many other modernist intellectuals, Krutch

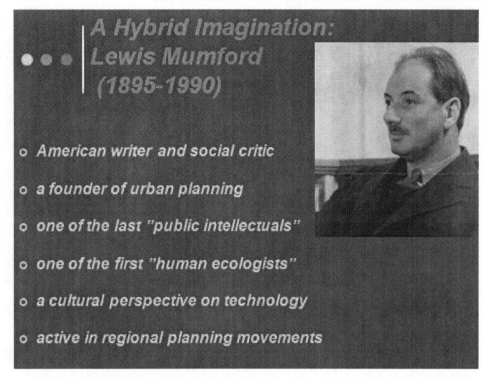

A Hybrid Imagination:
Lewis Mumford
(1895-1990)

o *American writer and social critic*

o *a founder of urban planning*

o *one of the last "public intellectuals"*

o *one of the first "human ecologists"*

o *a cultural perspective on technology*

o *active in regional planning movements*

had given vent to a kind of cultural despair in his book, and he would end up leaving New York and the intellectual life to move to the desert and become one of the early spokesmen for the emerging environmental movement.

In the many books that he would write about technology and cities, as well as about architecture and planning, Mumford combined ecology and history, geography and sociology, and an interest in both the past and the future in order to provide what might be termed a cultural assessment of modernity. Mumford's perspective resembled that of the academic field of human ecology, which developed in the interwar years among sociologists at the University of Chicago, who had been inspired by the progressive social work of Jane Addams, as well as among the "southern regionalists" such as Howard Odum at the University of North Carolina. For Mumford, technocratic thought represented primarily a failure of imagination. In glorifying the machine, the utopian side of man had developed into a sort of hubris, and with it, there was an urgent need for a cultural evaluation of the contemporary technological civilization [Jamison, A., 1998, Miller, D., 1989].

In *Technics and Civilization* (1934), Mumford drew both on his technical interests and his artistic inclinations to fashion a cultural interpretation of technology and science. He articulated a philosophy of "neotechnics" seeing in the new science-based innovations of the 20th century an organic, or what he termed organismic, attitude to reality that could supersede the mechanical

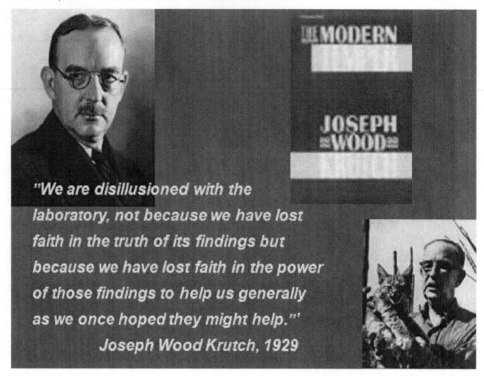

"We are disillusioned with the laboratory, not because we have lost faith in the truth of its findings but because we have lost faith in the power of those findings to help us generally as we once hoped they might help."

Joseph Wood Krutch, 1929

philosophy of the 18th and 19th centuries. The Wright brothers, for example, had succeeded where others had failed because they had closely observed nature:

> the final touch, necessary for stable flight, came when two bicycle mechanics, Orville and Wilbur Wright, studied the flight of birds, like the gull and the hawk, and discovered the function of warping the tips of the wings to achieve lateral stability. Further improvements in the design of airplanes have been associated, not merely with the mechanical perfection of the wings and the motors, but with the study of the flight of other types of bird, like the duck, and the movement of fish in water [Mumford, L., 1934, p. 251].

Technics and Civilization marks, in many respects, a turning point in the intellectual appropriation of technology and science in the 20th century. Where previous writers had either tended to glorify or denigrate the facts and artifacts of science and technology, Mumford sought to provide a more comprehensive mode of interpretation. He wanted to encompass both the positive and negative features, and develop what later came to be termed technology assessment. As the 1930s progressed, such technology assessment would become an integral part of the so-called New Deal era, with its programs of regional development, such as the Tennessee Valley Authority, where a series of hydroelectric dams were combined with other measures of regional planning to bring about a process of modernization in what had been a poor southern region of the country. At the time,

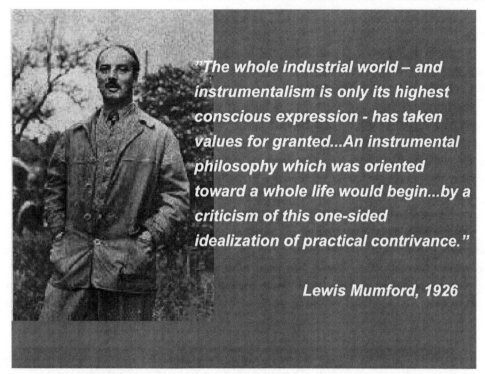

"The whole industrial world – and instrumentalism is only its highest conscious expression - has taken values for granted...An instrumental philosophy which was oriented toward a whole life would begin...by a criticism of this one-sided idealization of practical contrivance."

Lewis Mumford, 1926

the TVA was considered a model of how science and technology could be developed to benefit a rural community and not merely contribute to economic growth and industrial expansion. As David Lilienthal, the program's director, would describe it in a book, originally published in 1943, the TVA was "democracy on the march"[Lilienthal, D., 1966]. Interestingly enough, Lilienthal after the war would go to direct the Atomic Energy Commission until he resigned in 1950 over disagreements with the military.

5.6   ANTI-COLONIAL HYBRIDS: GANDHI AND TAGORE

The social and cultural movements that affected the relations between science, technology and society in the colonies of Western imperialism were different in many ways from those of the industrialized countries. But like the modernist movements in what is now termed the global North, the anti-colonial movements in the global South played a major role in the cultural appropriation of science and technology. In many of the independence struggles in Asia and Africa, the anti-colonial movements were led, in large measure, by movement intellectuals with hybrid imaginations, who sought to combine Western science and technology with non-Western values. These were often lawyers, doctors, and scientists and engineers who had been educated in Europe but who, in various ways, tried to mix what they had learned with the traditional beliefs that they had been brought up with.

As is well known, it was under the inspiration of Mohandas, later the Mahatma, Gandhi that the peoples of the Indian subcontinent were encouraged to revive traditional technical practices and combine them with Western ways to create an alternative kind of modernization. Gandhi had studied law in Britain, and there he had become acquainted with Western science and technology. Even more significantly, he had learned about Western traditions of cultural criticism and social movements. In particular, Gandhi had read the writings of Henry David Thoreau, who, in addition to his ecological interests had also been an active participant in the abolitionist movement and written the famous pamphlet "civil disobedience" to justify his refusal to pay taxes to an American government which sanctioned slavery [Dalton, D., 1993].

Throughout his life, Gandhi mixed the West and the East in a series of what he called "experiments with truth" (the title of his autobiography), and in the preface, he presents himself as something of an experimental scientist in the way he went about forming his ideas about independence [Gandhi, M.K., 1927]. The ideas of non-violence came directly from Thoreau's concept of civil disobedience, and his ideas about science and technology were infused with the ideas of romanticism and populism, which he combined with Indian traditional beliefs and practices.

A Hybrid Imagination:
M.K. (Mahatma) Gandhi (1869-1948)

"Just as matter displaced becomes dirt,
Reason misplaced becomes lunacy."

Gandhi was not alone in his attempts to develop alternative approaches to science and technology in pursuit of modernization. On the one hand, his strategy differed from the pro-Western nationalist leaders, such as Nehru, who stressed the importance of adopting western science and technology as the only way out of poverty and backwardness. But, on the other hand, Gandhi's hybrid imagination was different from that of artists and writers, such as Rabindranath Tagore, the first Asian to win the Nobel Prize. Gandhi's life-long debate with Tagore over the value of Western technology and science and, more specifically, over the value of spinning one's own cloth, is one of the defining tensions of modern India [Sen, A., 2005].

For Tagore, Gandhi's political genius and not least his great skills in mobilizing popular support were of questionable value when applied to matters of education and knowledge making. He criticized Gandhi's glorification of tradition, especially in relation to artistic and cultural expression. Tagore wrote popular science books, as well as literature, and he was an artisan craftsman as well as a painter and sculptor. Throughout his life, he argued for the importance of creative combinations of western and eastern approaches to modernization. He established schools in which both the indigenous, or traditional forms of art and knowledge, as well as western science and technology were taught. He communicated with western scientists and engineers, and when he received the Nobel Prize, he toured Europe and was well-received by the artists and writers of the modernist movement as one of them, as well as meeting with Einstein and other scientists [Dutta, K. and Robinson, A., 1996].

A Hybrid Imagination:
Rabindranath Tagore
(1861-1941)

"What I object to is the artificial arrangement by which the foreign education tends to occupy all the space of our national mind and thus kills, or hampers, the great opportunity for the creation of new thought by a new combination of truths..."

If Gandhi was too traditional for Nehru and Tagore, he was not traditional enough for Ananda Coomaraswamy and many of the other independence leaders. Ashis Nandy has contrasted Gandhi's "critical traditionalism" to the more absolute glorification of tradition represented by the art historian and Buddhist scholar Coomaraswamy, who served for many years as curator of Asian art at the Boston Museum of Fine Arts after taking part in the independence movement in Ceylon, now Sri Lanka. Where Gandhi made use of Indian traditions in an open-ended, reflective way, Coomaraswamy's

> tradition remains homogeneous and undifferentiated from the point of view of man-made suffering… Today, with the renewed interest in cultural visions, one has to be aware that commitment to traditions, too, can objectify by drawing a line between a culture and those who live by that culture, by setting up some as the true interpreters of a culture and the others as falsifiers, and by trying to defend the core of a culture from its periphery [Nandy, A., 1987, p. 121, 122].

While Gandhi was critical of Western technology and science, he was also highly critical of Indian traditions and, in particular, of the religious warfare that was so endemic to the subcontinent. It was the lack of morality, the lack of idealism in both Western and non-Western civilization that Gandhi objected to, and his own hybrid identity was a unique, not to say idiosyncratic mixture of the West and the East.

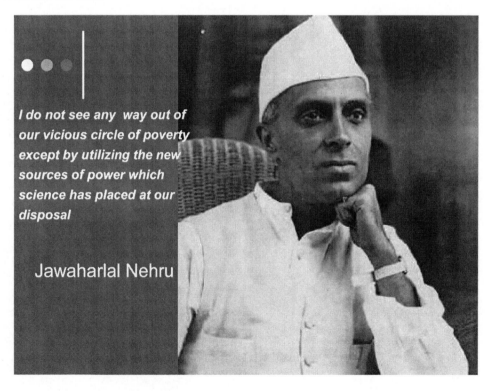

I do not see any way out of our vicious circle of poverty except by utilizing the new sources of power which science has placed at our disposal

Jawaharlal Nehru

The double nature of Gandhi's critique is important in understanding the subsequent dilemmas of Indian development. Unlike the leaders of most other independence movements in non-Western societies, Gandhi sought to combine independence with truth and love, that is, *swaraj* (self-rule) with what he called *satyagraha*, a word he made up by combining the Sanskrit words *satya* (truth) and *agraha* (holding firmly) [Dalton, D., 1993, p. 9].

India did not follow Gandhi's lead in the first two decades after independence. Instead, under the leadership of Nehru, ambitious efforts were made to bring western technology and science into Indian society. For Nehru, Indian civilization, with its superstitions and religious strife, was in need of radical change, and his governments did their utmost to develop both scientific institutions and a popular understanding and appreciation for science. From the late 1940s, scientific and technological research were organized roughly along the lines of the Soviet model, with central planning and strong state control over priorities and orientation [Visvanathan, S., 1984].

5.7   FROM MODERNISM TO MODERNIZATION

From the 1930s onward, there was a parting of the ways among modernists throughout the world. As has been the case with social and cultural movements, both before and since, modernism and anti-colonialism were victims of their own success. Modernist approaches in art and architecture and, not least industrial design, became extremely popular and commercially successful, and as the well-known stories of Frank Lloyd Wright and Pablo Picasso and so many others amply illustrate, fame and fortune induced more than a slight tendency to hubris on the part of many a modernist.

An important factor in the success story was also the establishment of new modernist institutions and buildings and, in some places, entire city areas that were subjected to modernist renewal. Particularly important were the museums and galleries and workshops for the display and sale of modern art, which helped bring modernism into the "mainstream" of mass society and popular culture. The Museum of Modern Art in New York, founded in 1929, was one of the first important modernist institutions, at one and the same time, a place for viewing the paintings and sculptures and films of the modernist movement, but also a new kind of art work itself, which would be emulated later by Frank Lloyd Wright and Frank Gehry in their designs for the Guggenheim Museums in New York and Bilbao. The museum of modern art would become an emblem of modernity, a fixture of the 20th century city a center for public education and entertainment, cultural expression, as well as a commercial tourist attraction.

There was also a range of new educational institutions that were established, for the most part, outside of the traditional universities. New schools for designers and architects – as well as for other modernist intellectuals, such as the New School for Social Research in New York - were established throughout Europe and North America, and many artisan studios and architectural firms came to take on educational functions, as was the case with Frank Lloyd Wright's adventurous efforts in Wisconsin and Arizona. As at the Bauhaus and the Art Institute of Chicago, where many of the Bauhaus staff had moved after the Nazi ascension to power, the teaching staff at many of these new educational institutions included working artists, architects, planners and engineers, as well

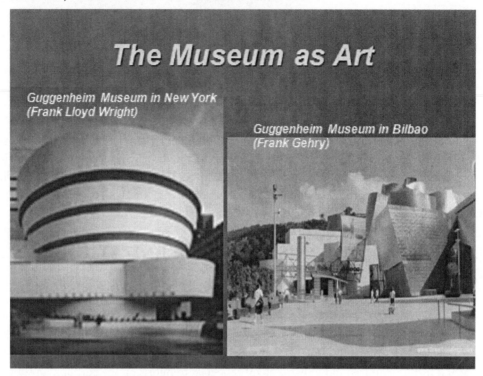

The Museum as Art

*Guggenheim Museum in New York
(Frank Lloyd Wright)*

*Guggenheim Museum in Bilbao
(Frank Gehry)*

as various academic scientists, and the places themselves, in their lecture halls and theaters, their cinemas and galleries, became in many parts of the world public centers for viewing and exhibiting and sometimes even critically reflecting upon the pros and cons of modern civilization.

Other sites, or cultural contexts, were no less important for what might be called the institutionalization of modernism. In Barcelona, many private and public buildings, as well as parks and squares were reconstructed by Antonio Gaudi according to his particular version of modernist architecture and design. His still unfinished cathedral, in particular, became an important tourist attraction as art mixed with urban planning to revive and renew urban centers. As decorators of streets and highways and, not least, as builders and designers of the service facilities for the new technological artifacts – the gas stations and the department stores and eventually the malls of modern society – the modernist movement helped shape the physical landscape of 20th century civilization.

In Sweden and the other Nordic countries, the development of design and modern art were important for national survival itself, and industrial design and architecture became crucial components in the belated process of industrial transformation, as well as contributing to the formation of national identity. The countries of northern Europe became the quintessential modernist places, not least because of the architecture of Eero Saarinen, Alvar Aalto, Arne Jacobsen, and so many others. Nordic design drew on the aesthetic principles of William Morris and the arts and crafts movement of the 19th century, but those principles and practices were mixed with Nordic traditions to create artifacts of natural simplicity and beauty that were also practical and functional.

If modernism was a victim of its own success, its demise as a movement was also the result of the increasingly polarized and unstable political climate of the 1930s and, not least, the rise of aggressive counter-movements of "anti-modernism" that took power in Germany, Spain, Italy, Japan and elsewhere. Also in the Soviet Union, the consolidation of Stalinism in the 1930s meant that the kind of experimental hybridity that had been so widespread in both the arts and the sciences in the 1920s was more or less brought to an end. As with the socialist movements of the 19th century, "residual" cultural formations were mobilized in active opposition to the modernist movement, and as the 1930s progressed, the movement had more or less fragmented into what we have termed the contending cultures of science and engineering: a market-oriented, or business culture, on the one side, and a state-oriented, or bureaucratic culture, on the other, with a few small countries, such as Sweden trying to carve out a third, or "middle way" in between.

5.8   THE MILITARIZATION OF MODERNITY

In the sciences, the overturning of the classical mechanics in both theory and practice that had begun with Einstein and Bohr eventually encouraged another kind of hubris, especially when the new theories were "tested" in laboratory experiments to attempt to extract new sources of energy from atoms and their nuclei. In the 1930s, Enrico Fermi and Leo Szilard succeeded in producing what came to be called atomic energy, by setting off a chain reaction in uranium atoms, a process that came to be called nuclear fission, as the nucleus in the atom literally split apart and exploded.

Their efforts would come to have more of an impact on human history than perhaps any other scientific or technological achievement in the 20th century. By a tragic accident of history, the making of atomic energy and the theoretical understanding that went with it came to be skewed, or misdirected to military purposes, due to the fact that many of the most active scientists were Jews, who had fled from Europe, and were terrified at the prospect of Germany developing an atomic bomb, which they knew was fully possible. When war broke out, Leo Szilard asked Albert Einstein, both of whom were Jews who had emigrated to the United States to escape the Nazis, to write to President Roosevelt and propose that the Americans build a bomb before the Germans managed to do it. It was thus that the pacifist Albert Einstein would be directly responsible for the most horrendous weapon ever built.

The scientific and technological development of atomic energy was a part of what in the 1940s became a worldwide militarization of science and technology. As the nations of the world went to war for the second time within the space of thirty years, all those who had identified with a modernist movement were forced to take sides and most of them took active part in the war effort, which in many ways pitted the modernist movement against the forces of anti-modernism. As we shall see in the following chapter, while the reconstruction of science and technology in the 1940s and 1950s would thus draw on modernist ideas and practices, it would do so in a particularly militarized fashion. In first the second world war, and then in the Cold War between capitalism and communism, much of the development of science and technology would take place in uniform, or at least under the sponsorship of the military. Both the transistor and the personal computer, and, indeed, the Internet itself would be developed under military auspices.

CHAPTER 6

Science, Technology and Globalization

Technology has become the great vehicle of *reification* – reification in its most mature and effective form. The social position of the individual and his relation to others appear not only to be determined by objective qualities and laws, but these qualities and laws seem to lose their mysterious and uncontrollable character; they appear as calculable manifestations of (scientific) rationality. The world tends to become the stuff of total administration, which absorbs even the administrators.

Herbert Marcuse, *One-Dimensional Man* (1964)

6.1 A NEW MODE OF SCIENCE AND TECHNOLOGY

In the decades that have followed the Second World War, science and technology have come to play ever more central roles in our societies, our economies, as well as in our personal lives. From the never ending stream of research-based technological apparatus that has become so essential for keeping us busy and happy and our countries "competitive" to the political disputes over climate change, genetically modified foods and environmental pollution, scientific facts and technological artifacts permeate our existence. They have infiltrated our languages, altered our behavior, and, perhaps most fundamentally, imposed their instrumental logic – their scientific and technological rationality – on the ways in which we interact and communicate with one another.

In contextual terms, there has been a transition during the second half of the 20th century, from a "mode" of knowledge production that was largely self-governed and thereby controlled by the knowledge producers themselves, into networks or systems of innovation that are sponsored by state and inter-state funding agencies and business firms. The borders that used to separate universities from the life-worlds of business and government have been transgressed in the vast array of science parks and university-industry collaborations that were initiated in the 1950s in California's Silicon Valley. Like the silicon, or microelectronic chip itself, Silicon Valley was born as one of the many "spin-offs" of the militarization of science and technology in the decades following the Second World War.

Transdisciplinarity, or "mode 2"

"Knowledge which emerges from a particular context of application with its own distinct theoretical structures, research methods and modes of practice but which may not be locatable on the prevailing disciplinary map."

Michael Gibbons et al, The New Production of Knowledge (1994)

In cognitive terms, the boundaries that had previously distinguished scientific research, or philosophical-theoretical knowledge from technological development, or practical-technical knowledge have been blurred, or transgressed in many, if not most fields of science and technology. In information and communication technologies, genetic and other biotechnologies, and, more recently, media engineering, cognitive science, nanotechnology, and synthetic biology, as well as in many other areas of contemporary science and technology, there is no clear line of demarcation between scientific "theory" and engineering "practice." For the most part, knowledge is made in temporary project groups and networks that bring university scientists and engineers together with corporate employees and government officials.

As such, the quality, or truth value of this transdisciplinary knowledge depends on the context in which the knowledge is made. Instead of solely evaluating their truth claims among themselves according to criteria that are internal to scientific and engineering fields, "quality control," as Gibbons et al. put it, "is additionally guided by a good deal of practical, societal, policy-related concerns, so that whatever knowledge is actually produced, the environment already structured by application or use will have to be taken into account"[Gibbons et al., 1994, p. 33].

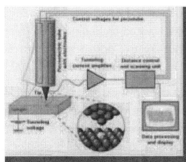

The Challenge of Technoscience

o *A blurring of discursive boundaries*
  * *between science and technology, nature and humanity*

o *A trespassing of institutional borders*
  * *between public and private, economic and academic*

o *A mixing of skills and competencies*
  * *between theoretical and practical knowledge*

In quantitative terms, the resources devoted to science and technology – both human and financial – have increased enormously during the past seventy-five years, beginning with the large-scale mobilization of scientists and engineers for the war effort itself and in the massive scaling up that took place throughout the world in the 1940s and 1950s, especially in the development of atomic energy. Particularly in the United States and the Soviet Union, but in most of the other countries in their spheres of influence as well, science and technology became integral parts of the cold war, not least in regard to the so-called "arms race:" the making of science-based weapons of mass destruction. While the U.S. and the Soviet Union, as the main protagonists, came to devote well over half of their scientific and technological resources to military purposes, they were still able to lead the world in most areas of research and development in the immediate postwar decades. But eventually, other countries that had proportionately smaller military budgets – first Japan and then South Korea and the smaller countries of Southeast Asia, and more recently China and India – were able to surpass them in many fields, especially in the area of information and communication technology.

According to the dominant story-line, it has been the so-called globalization of economic production, consumption and commerce, driven by the widespread diffusion and application of information and communication technologies that has been the main factor behind the changing relations between science, technology and the broader society and economy. Together with genetic engineering and other biotechnologies, these "high," or advanced technologies are seen to represent a technological revolution, ushering in a new wave of creative destruction that resembles the Schumpeterian long waves of the past. In the words of Freeman, C. and Louçá, F. [2003, p. 301]:

> Even those who have disputed the revolutionary character of earlier waves of technical change often have little difficulty in accepting that a vast technological revolution is taking place, based on the electronic computer, software, microelectronics, the Internet and mobile telephones... From a very much smaller base and on a much smaller scale, bio-technology was also growing very rapidly in the closing decades of the twentieth century. In one sense, it too is a very special form of information technology and it is interacting increasingly with computer technology.

In the 21st century, nanotechnology and cognitive science, and, most recently, synthetic biology, have come to be seen by many, if not most people involved with scientific and technological development, to provide the basis for an entirely different "knowledge-based" economy and society. This is the perspective, or story-line, that guides policy-makers throughout the world as they make their priorities and their decisions about which kinds of science and technology to support.

A report to the European Commission, "Converging Technologies – Shaping the Future of European Societies" by a working group with the name, "Foresighting the New Technology Wave" presents a typical example of the kinds of stories that are told about science and technology in policy circles [HLEG, 2004]. As the group's title implies, the report foresees a new wave of science and technology-based progress triggered by the so-called converging technologies. Nanotechnology, biotechnology and information technology – "nano, bio, info" – are connected to a number of other sciences and technologies that are characterized as "cogno, socio, anthro, philo, geo, eco, urbo, orbo, macro, macro" in the overall concept of converging technologies, or CTs for short.

CTs, it is suggested, have four major characteristics. They form an "invisible technical infrastructure for human action – analogous to the visible infrastructure provided by buildings and cities." They have an "unlimited reach:" the experts tell us that "it would appear that nothing can escape the reach of CTs and that the mind, social interactions, communication, and emotional states can all be engineered." This leads to the third characteristic, "engineering the Mind and the Body," and in the report, it is claimed that "humans may be drawn to surrender more and more of their freedom and responsibility to a mechanical world that acts for them." The final characteristic of CTs is their specificity: "research on the interface between nano and biotechnology allows for the targeted delivery of designer pharmaceuticals that are tailored to an individual's genome in order to affect a cure without side effects. More generally, the convergence of enabling technologies and knowledge systems can be geared to address very specific tasks." And while there may be some undesirable side-effects of these converging technologies, the experts make it clear in their report that there is no turning back. If Europe is to compete successfully in the global marketplace, then the European Commission needs to expand its support, not least for the newest of the converging technologies, nanotechnology, which has come to be seen by many policy-makers as the strategically most important of the new technoscience fields. In the words of the European Commissioner for science and research, Janez Potočnik:

> Nanotechnology is an area which has highly promising prospects for turning funda-mental research into successful innovations. Not only to boost the competitiveness of our industry but also to create new products that will make positive changes in the lives of our citizens, be it in medicine, environment, electronics or any other fieldtechnol-ogy [EU Nanotechnology, 2011].

Most of the popular and academic discussion about science and technology in the contempo-rary world follows this dominant story-line of economic innovation, while the products of high-tech industry provide us with the wires with which we connect ourselves to one another in the networks of our lives, both at work and at play. Globalization is seen as an unstoppable process of economic innovation and entrepreneurial innovators who can bring science and technology to market and create new companies, and branches of industry are seen as the harbingers of a new age and makers of a new, virtual reality. As Castells, M. [1996] has put it:

> Toward the end of the second millennium of the Christian era several events of historical significance transformed the social landscape of human life. A technological revolution, centered around information technologies, began to reshape, at accelerated pace, the ma-terial basis of society... *Our societies are increasingly structured around a bipolar opposition between the Net and the self.*

The social construction story-line, on the other hand, which has been central to much of the academic study of science and technology among social scientists and historians focuses on the changes in knowledge production itself, that is, the ways in which scientists and engineers actually make, or construct new "facts" and "artifacts." The shift to technoscience is seen primarily as a transformation, in the influential words of Latour, B. [1996], from Science to Research:

> Looking for an expression that could capture the change that has occurred in the last century and a half in the relation between science and society, I can find no better way than to say that we have shifted from Science to Research. Science is certainty; Research is uncertainty. Science is supposed to be cold, straight and detached; Research is warm, involving and risky. Science puts an end to the vagaries of human disputes; Research fuels controversies by more controversies. Science produces objectivity by escaping as much as possible from the shackles of ideology, passions and emotions; Research feeds on all of those as so many handles to render familiar new objects of enquiry.

With the opening of new realms of reality for researchers to investigate – virtual, molecular, sub-atomic, and nanoscale – the traditional boundaries between nature and society, or what Latour calls the "modern constitution" separating humans from non-humans, have been blurred beyond recognition. According to the story-line of construction, this means that scientists and engineers need to change their professional identities and see themselves as constructors, not merely of natural and

> ### ● ● ● | *Contending Responses to Technoscience*
>
> o *The dominant, or business strategy ("mode 2"):*
>   *innovation, entrepreneurship, transdisciplinarity*
>
> o *The residual, or professional strategy ("mode 1"):*
>   *construction, expertise, (sub)disciplinarity*
>
> o *An emerging, or hybrid strategy ("mode 3"):*
>   *appropriation, empowerment, cross-disciplinarity*

technical things but of "actor-networks" as well. The task for scientists and engineers is to "reassemble the social" in new configurations or assemblages of humans and non-humans, as Latour, B. [2005] puts it in the title of a recent book.

In this chapter, we want to show how the changing relations between science, technology and society during the past seventy-five years have been influenced by social and cultural movements and, more specifically, by the "new social movements" that emerged in the 1960s and 1970s. These movements – of environmentalism, women's liberation, and anti-imperialism, as well as the so-called "counter-culture" – represented, among other things, a global protest against the kinds of science and technology that had been developed during the Second World War and in the "cold war" that followed. They protested against the militarization of science and technology, as well as its dehumanizing and inhuman effects. In their various ways, the social and cultural movements of the 1960s and 1970s sought to liberate science and technology from the military-industrial complex and other bureaucratic contexts.

Their overarching "cosmological" ambition was to relate knowledge making – in a phrase that was often used in the 1970s – to the satisfaction of "basic human needs." In their efforts to foster more humane, or appropriate technologies, they produced innovations in a wide range of scientific and engineering fields, from personal computers to wind energy, from organic agriculture

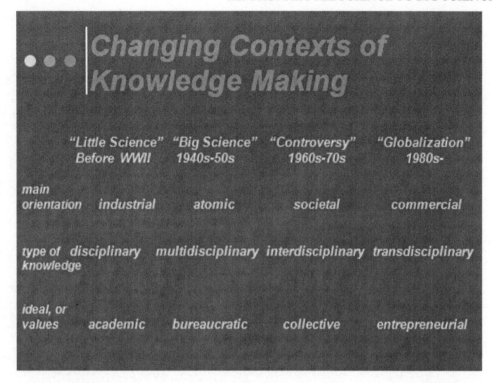

to ecological design, and from reproductive technologies to genetic engineering. Like the social and cultural movements that we have discussed in previous chapters, the movements of the 1960s and 1970s played a major role in the creative reconstruction of science and technology that has taken place since the 1980s.

6.2   FROM LITTLE SCIENCE TO BIG SCIENCE

During World War Two, scientists and engineers were supported by society to an unprecedented extent in order to produce more effective weapons, as well as provide expert advice and secret intelligence that could contribute to the war effort. In addition to the atomic bomb, or Manhattan project, radar, computing, synthetic chemistry, rockets and ballistic missiles were all areas to which substantial research and development efforts were devoted during the Second World War. The wartime mobilization of technology and science would initiate new kinds of relations between science, technology and society. From the 1940s through the 1960s, internally-driven approaches to the production of knowledge, based on disciplinary identities and academic values, came to be complemented by externally-imposed institutional forms and bureaucratic values. As Price, D.S. [1963] put it, when he summarized his statistical analysis of rates of increase in money and manpower devoted to research and development through the 1950s, "little science" had given way to "big science."

> ••• | **A Fourth Cycle**
>
> o *the age of "big science" (ca 1930-1980)*
>
> o *Atomic energy, petrochemicals, and computers*
>
> o *Technologies of scientification*
>
> o *The rise of transnational corporations*
>
> o *Social and cultural movements:*
>
> > • *civil rights and "ban the bomb"*
> >
> > • *environmentalism, feminism and postmodernism*

Largely because of the decisive role that they had played in the war effort, the social status and prestige of scientists and engineers changed significantly when the war had ended. The funding and organization of science and technology became a new area of concern for national governments, in addition to the relatively few private corporations that had provided "external" support before the war. In the words of MIT engineering professor Vannevar Bush, who, after serving as a wartime government adviser, was asked by President Roosevelt to suggest how the US government should best deal with this new role, science was characterized as the new "frontier." And the frontier that was science, unlike the frontier of the old West, was considered to be "endless."

The Bush report, *Science, the Endless Frontier*, discussed how the experiences of mobilizing science and technology for the purposes of war could be applied to peaceful or civilian purposes. On a discursive level, Bush and his counterparts in other countries fashioned a strategic narrative: science was characterized as a crucial resource in what was soon to be perceived as an international power struggle. What had been achieved in wartime should now be transferred to the marketplace and in "international relations" (which became a scientific subject of its own, along with strategic studies after the war). For strategic reasons, substantially larger amounts of public funding ought

● ● ● **From Little Science to Big Science**

o *result of use of science in WW2*

o *change in size and scale*

o *strategic orientation, external control*

o *university-government collaboration*

o *bureaucratic norm, or value system*

o *new role for the state: "science policy"*

to be channeled to science and engineering, as well as to higher education, or, as Bush put it, "the Federal Government should accept new responsibilities for promoting the creation of new scientific knowledge and the development of scientific talent in our youth"[Bush, V., 1960, p. 31].

In return for giving scientists and engineers vastly increased amounts of money, Bush presented a vision of unimagined prosperity and material abundance. At the institutional level, a range of new research councils and other governmental bodies were created throughout the world (the Bush report led to the establishment of the National Science Foundation in the U.S.). Major research and development – or R&D – institutions were also set up in the immediate aftermath of the war, particularly in order to develop "civilian uses" of atomic energy. In Europe, a number of national governments joined together to create CERN, a center for nuclear energy research in Switzerland. These state-supported facilities complemented those already established by the military and provided a new set of large-scale, multidisciplinary sites for carrying out research and development activities that neither traditional universities nor private corporations could afford. It was at one such institution – at Oak Ridge, Tennessee – that the director, Alvin Weinberg coined the term, Big Science, as a way to distinguish the kind of knowledge produced at such places from the "little science" of the past [Weinberg, A., 1967].

Combining traditional scientific norms and professional engineering values with the demands of large-scale bureaucracies proved to be easier said than done, however, and there was a good deal of discussion as the 1950s progressed about the resultant cultural tensions, from C.P. Snow's famous lecture on the division of society into "two cultures" – a scientific-technical and a literary-artistic – to President Eisenhower's concerns, which he expressed on leaving office in 1960, about the growing power of the "military-industrial complex" over society. The new relations between science, technology and society were countered by scientists such as Albert Einstein and Leo Szilard, who in the course of the 1950s, organized meetings between scientists in the West and in the Soviet bloc to discuss disarmament issues and world peace, such as the so-called Pugwash conferences. Szilard, whose work in the 1930s with Enrico Fermi had been crucially important for the development of atomic energy, left physics altogether to devote his time to biological and social research and, for the most part, nuclear disarmament. Many were the scientists and engineers who contended that the values of internationalism, academic freedom, and what Karl Popper more generally termed the "open society," were threatened by the new kinds of relations that were developing between science, technology and society, and, not least, by the domination over science and technology by the military [Jamison, A. and Eyerman, R., 1994].

Big Science as Hubris

As the 1950s wore on, it became clear to politicians, as well as to the general public that, for all the money being spent on science, the United States was no longer the unquestioned global leader in all scientific and technological fields. In other capitalist countries, such as Japan and (West) Germany, as well as in the Soviet Union, the state did not merely support basic research but industrial technological development as well. The shock of the *Sputnik* satellite, which the Soviets sent into orbit in 1957, served to initiate important changes both in the discourses of science and technology policy as well as in the practical and institutional dimensions of knowledge production. As economists began to explore the innovation process, as it started to be called, it became clear to many that technological innovations were not merely a matter of applying the results of "basic science," as Bush had implied in his report after the war but required a more sophisticated understanding of firm strategies and the dynamics of technological development [Freeman, C., 1974].

The changing relations between science, technology and society had a fundamental influence on the theory of science, as philosophers and historians came to debate the dynamics of scientific growth. On the one side were philosophers, led by Karl Popper, who contended that science grew continuously and cumulatively, and on the other side was the physicist-turned-historian Thomas Kuhn, who recognized the social conditioning of scientific knowledge and presented science as a discontinuous process, a series of paradigm shifts and "scientific revolutions"[Kuhn, T., 1962].

● ● ● | *Thomas Kuhn's Revolution*

- o *a collective, "big science" model of science*

- o *research guided by paradigms, or disciplinary matrices*

- o *"normal" science as a form of puzzle-solving*

- o *disrupted by periodic revolutions: paradigm conflicts*

- o *science is a discontinuous process*

In his extremely influential book, *The Structure of Scientific Revolutions*, Kuhn distinguished between "normal science" which he characterized as a kind of puzzle-solving, from the "revolutionary" science that led to major shifts in the paradigms or what he later termed the disciplinary matrices that scientists followed. Throughout the history of science, the revolutions had not come without a struggle, as scientists sought to hold onto their traditional, or habitual ways of doing research. It was only when "anomalies" occurred which the paradigm could not explain – such as the discovery of radiation and the measurement of the speed of light, which challenged the classical paradigms of physics – that a period of scientific revolution took place and a new paradigm emerged – such as that which had taken place in the 1910s and 1920s and led to relativity theory and quantum mechanics. In scientific growth, there was thus an underlying tension between tradition and revolution. And in the cultural climate of the 1960s, Kuhn's book became a part of the widespread questioning as to whether the new, more bureaucratic – and militarized – contexts in which science and technology was carried out constrained the kind of original thought that was necessary for scientific revolutions.

6.3   A PERIOD OF QUESTIONING AND CRITIQUE

Throughout the world, the 1960s would be marked by a wide-ranging social and cultural critique of the priorities and policies of the big science era, which had already begun in the 1950s in regard to the continuing expansion of the nuclear arms race following the detonation of the first hydrogen bomb in 1954. Public demonstrations to "ban the bomb" were held in the late 1950s, with the philosopher-turned-activist Bertrand Russell playing one of the leading roles.

In the course of the 1960s, the questioning would come to encompass much more than nuclear weapons, taking on big science in general and the discourses, institutions and practices that went with it. We have previously mentioned the moral, or spiritual critique that was voiced by religious leaders, such as Martin Luther King in the United States. What King termed the poverty of the spirit was part of a more general concern with the lack of morality and the violations of human or civil rights that was so widespread in the contemporary world. The development of science and technology had turned citizens into consumers, and as a result, many contended that there was a need to bring a new kind of ethical or humanitarian concern into the making of science and technology (cf. Mitcham, C. and Muñoz, D., 2010).

Another kind of critique concerned the impact that scientific and technological development was having on nature or what came to be referred to in the 1960s as the natural environment. The environmental movement, which we will discuss in more detail in the following chapter, was set off by the book, *Silent Spring*, written by the biologist-turned-science writer, Rachel Carson.

While conservationists had been discussing the consequences that science-based economic development was having on plants and animals, it would be Carson's book, with its detailed exposé of the health and environmental implications of one particular, widely-used chemical in agriculture, the insecticide DDT – which had been invented during the Second World War – that would bring the environmental cause to public attention. It would also usher in a more activist and radical approach to environmental politics than had been characteristic of the older conservation societies which had

> ● ● ● | **Critiques of Big Science in the 1960s**
>
> ○ **moral, or spiritual (e.g. Martin Luther King)**
>   - *against injustice, "poverty of the spirit"*
>   - *for a new morality, or sense of justice*
>
> ○ **ecological, or internal (e.g. Rachel Carson)**
>   - *against reductionism, "the abuse of the planet"*
>   - *for a new, environmental science*
>
> ○ **humanist, or cultural (e.g. Lewis Mumford)**
>   - *against hubris, "the myth of the machine"*
>   - *for an appropriate technology*

been established in the late 19th and early 20th centuries and tended to be located on the conservative side of the political spectrum.

What Carson and other environmentalists argued was that a full-fledged crisis was in the offing if science and technology were not changed into more environmentally-friendly or ecological directions (Carson 1962; cf. Jamison, A. and Eyerman, R., 1994). Many of the new kinds of science-based products that had been produced in the immediate postwar era, especially the synthetic chemicals that were used in agriculture and food production, and many health and household products, as well, could not be broken down and recycled in nature as could the products they replaced, and thus served to destroy the natural environment, as well as affecting human health.

In addition to the environmental debate, there also emerged in the 1960s a more general questioning of the ways in which the broader society had been affected by the priorities of the big science era. The increasingly visible and horrific uses of science and technology in the war in Vietnam as well as the more general lack of a broader social responsibility in the ways that students were being educated brought on a wave of student revolts in the second half of the 1960s. Humanist scholars and philosophers, such as Hannah Arendt and Herbert Marcuse, who had fled from Nazism, saw in the scientific-technological state a new form of authoritarianism and wrote influential books about

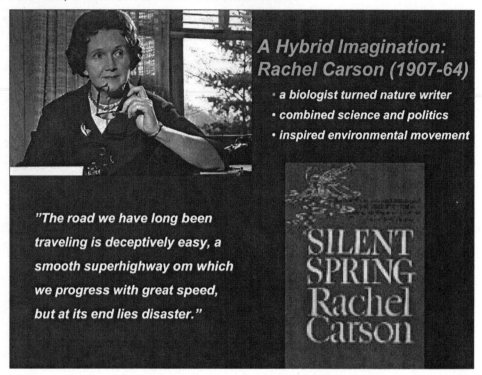

A Hybrid Imagination:
Rachel Carson (1907-64)

• *a biologist turned nature writer*
• *combined science and politics*
• *inspired environmental movement*

"The road we have long been traveling is deceptively easy, a smooth superhighway om which we progress with great speed, but at its end lies disaster."

SILENT SPRING Rachel Carson

what Arendt called the "human condition" and what Marcuse called "technological rationality" and one-dimensional thought [Jamison, A. and Eyerman, R., 1994].

One of those who was most active in this humanist critique of big science was Lewis Mumford, who throughout the postwar era kept on writing books, as well as numerous articles in such popular magazines as the *New Yorker* about urban and regional planning, architecture, as well as technological and scientific development. In the 1960s, Mumford published two volumes on the "myth of the machine," bringing up to date the pioneering analysis of the relations between science, technology and society that he had made in the 1930s in *Technics and Civilization*, which we discussed in the previous chapter.

For Mumford, the development of science and technology in the postwar era had been dominated by what he termed the "megamachine" with its close ties to the military and the so-called military-industrial complex. He contrasted the "authoritarian" technics that had become so dominant in the postwar era – the "pentagon of power," as he called it in the title of the second book in his series – to what he termed "democratic" technics, which, throughout history, had represented the ways that people had dealt with their myriad problems of survival and development. Mumford's books were to be one of the sources of inspiration for the various attempts in the 1970s to develop more "appropriate" technologies in what became a significant process of collective learning in the social and cultural movements that developed.

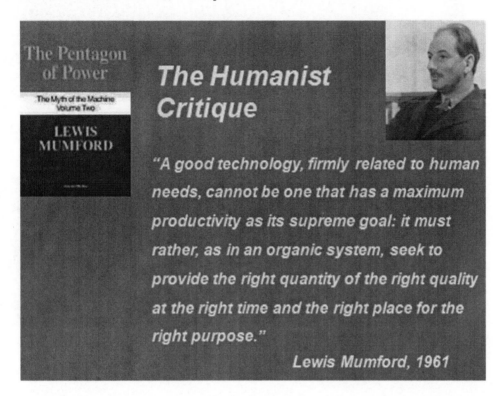

The Humanist Critique

"A good technology, firmly related to human needs, cannot be one that has a maximum productivity as its supreme goal: it must rather, as in an organic system, seek to provide the right quantity of the right quality at the right time and the right place for the right purpose."

Lewis Mumford, 1961

As science and technology had become ever more integrated into the economy and the state, a gap had emerged, not least in education, between what the British chemist-turned-novelist C.P. Snow termed the "two cultures" in a famous lecture in 1959. Snow's argument, which was echoed by many others throughout the world in the course of the 1960s was that, both in education as well in the broader culture, scientists and engineers, on the one side, and humanists and writers, on the other, had come to form separate cultural identities in the postwar era. Education and communication both in the professional and popular media had become polarized and overly specialized, and there was a need for both sides to know more about what the other was doing.

As a "new left" emerged as part of the student revolts, alternative ideas about science, technology and society were promulgated both among scientists and engineers, as well as among concerned citizens and even policy-makers. In several countries, societies for social responsibility in science were established, and in many national governments, as well as the EEC and OECD, new kinds of socially-oriented science and technology policies began to be formulated. In 1971, an OECD committee, headed by Harvard engineering professor Harvey Brooks produced the report, *Science, Growth and Society*, which was one of the most explicit attempts to respond to the questioning and

A Hybrid Imagination:
Charles Percy Snow (1905-1980)

• *chemist-turned-novelist*
• *active in science policy*
• *wrote literature about scientists*

The Two
Cultures
C. P. SNOW

Technology is [...] a queer thing. It brings you gifts with one hand, and stabs you in the back with the other

debate of the 1960s. Rather than defining the task of the government primarily in terms of national security and military defense, the report contended that the state should take on a much broader role if society were to benefit from science and technology. In the following years, many countries throughout the world would create new agencies to support research in socially-relevant fields of science and technology, in such fields as environmental protection, social welfare, public health, energy production, and transportation, as well as technology assessment, following the establishment of the Office of Technology Assessment in the U.S. in 1974. In general terms, there was a widening of focus in governmental activity, so that more policy sectors were given the capability to support and use scientific research and technological development.

Another outcome of the debates of the 1960s was the emergence of teaching and research programs in science, technology and society (STS) at universities throughout the world, to a large extent, to try to bridge the "two cultures" gap. The idea was to offer instruction about the social and cultural contexts of science and technology, as well as to provide meeting places for natural scientists, engineers, social scientists and humanists for discussion seminars and workshops and eventually for carrying out research projects together. The field of STS, at least at the beginning, was part of a more general interest within universities to foster interdisciplinary studies in a number of new fields.

In the course of the 1970s, courses and entire departments in women's studies, peace studies, development studies and environmental studies – and, in the United States, African-American studies – as well as STS, would be established at many universities. A number of new universities were also created, often based on "student-centered" approaches to education that tried to transform the critical energy of the social movements of the times into more constructive directions. In Denmark, the new universities in Roskilde (1972) and Aalborg (1974) have ever since combined what has been called problem and project-based learning, as opposed to the more traditional "book learning" that characterized the older universities.

When applied to science and engineering, problem-based learning proved to be particularly effective as a way to connect university scientists, engineers and their students more closely to the problems in the broader society and to help cultivate the sorts of communicative, managerial and design skills that scientists and engineers would need if they were to be able to carry out their research and development work in a socially responsible manner.

In the course of the 1970s, there were also a number of centers set up outside the universities for appropriate, alternative, small-scale and/or intermediate technologies, putting into practice the ideas that were propagated in such books as *Small is Beautiful*, by E.F. Schumacher, an economist who had worked on development projects in India as well as for the British Coal Board. In the United States, a group of scientists and engineers left MIT to set up a "New Alchemy Institute" on Cape Cod, and for several years they held courses and developed research projects combining organic agriculture, renewable energy, and other "ecological technologies." In the general spirit of "liberation" that filled the air at the time, many scientists and engineers throughout the world, but perhaps especially in the so-called third world, sought to find ways to connect their scientific knowledge and technological skills to basic human needs. This was the expression used in many United Nations agencies in their activities and programs, as well as at the UN Conference on Science, Technology and Development, that was held in 1979 as part of the efforts on the part of developing countries to establish a "new international economic order."

6.4   FROM COUNTERCULTURE TO THE INFORMATION AGE

Another extremely relevant part of the social and cultural movements of the 1960s and 1970s for the future development of science and technology was the so-called "counter-culture" with its quest for personal liberation and the expansion of human consciousness [Roszak, T., 1969]. Like the other movements that we have discussed in this book, the counterculture is difficult to define, if it can be defined at all. We remember it mainly through its singers and its songs – Bob Dylan and "Like a Rolling Stone," Jim Morrison and "Light My Fire," Jimi Hendrix and "Purple Haze," Janis Joplin and "A Piece of My Heart" – and its cultural climax at Woodstock in 1969 and the battle for a People's Park in Berkeley, California. We all know that it was a significant part of what the sixties were all about, but what contribution did it actually make for the subsequent creative reconstruction of science and technology?

At the time, it was seen to represent a search for something beyond what Theodore Roszak referred to as the "myth of objective consciousness" toward the end of his book, *The Making of the Counter Culture*, that was published in 1969:

> If there is one especially striking feature of the new radicalism we have been survey-ing, it is the cleavage that exists between it and the radicalism of previous generations where the subjects of science and technology are concerned. To the older collectivist ideologies… science was almost invariably seen as an undisputed social good, because it had become so intimately related in the popular mind…to the technological progress that promised security and affluence. It was not foreseen even by gifted social critics that the impersonal, large-scale social processes to which technological progress gives rise – in economics, in politics, in education, in every aspect of life – generate their own characteristic problems [Roszak, T., 1969, p. 205].

In his book, Roszak presented the intellectual sources of the counter-culture in writers such as Hermann Hesse and Allen Ginsberg, psychologists such as R.D. Laing and Norman Brown, and philosophers such as Jacques Ellul and Herbert Marcuse in their very different critiques of what he termed the "technocracy" or what Marcuse in his *One Dimensional Man* had termed the reification of technology in contemporary society. "In the technocracy," Roszak wrote, "everything aspires to become purely technical, the subject of professional attention. The technocracy is therefore the regime of experts – or of those who can employ the experts [Roszak, T., 1969, p. 7]."

In his book, Roszak discussed the "journeys to the East – and beyond" that many took at the time with the help of Indian mystics and the Harvard professor Timothy Leary, who "dropped out" of the academic world to devote his life to getting people to "turn on" to hallucinatory drugs. He characterized the counter culture, as he would later do in a number of subsequent books, as a wide-ranging attempt to develop alternative forms of consciousness that were both ecological, holistic, and spiritual. In *Where the Wasteland Ends*, from 1973, he resuscitated the romantic poets of the 19th century, with their revolt against the instrumental and mechanical science of Bacon and Newton, and called for a new kind of science and technology that also could include feelings and emotions, and respect for the natural world.

Much of the counter-culture was, of course, not particularly interested in technology and science, aside from its uses for "blowin' your mind" through music and psychedelic drugs, but there were some who would turn the interest in personal liberation and alternative consciousness into scientific and technological development. Particularly among science and engineering students at Stanford and the University of California in Berkeley, as well as at MIT and other technological universities, something that came to be called a "hacker ethic" emerged in the 1960s, which would have major implications for the creative reconstruction of science and technology in the decades to come [Markoff, J., 2005].

As Levy, S. [1984] has characterized it in his book, *Hackers: Heroes of the Computer Revolution*, the hacker ethic had six central planks, or elements:

Access to computers – and anything which might teach you something about the way the world works – should be unlimited and total. Always yield to the Hands-On Imperative!

All information should be free.

Mistrust authority – promote decentralization.

Hackers should be judged by their hacking, not bogus criteria such as degrees, age, race, or position.

You can create art and beauty on a computer.

Computers can change your life for the better.

Steven Levy describes in his book how the hacker ethic had a significant influence on people like Steve Wozniak and Bill Gates, as they came up with their epochal inventions in the 1970s, the personal computer and computer software language. More recently, Turner, F. [2006] has shown in his book, *From Counterculture to Cyberculture: Stewart Brand, the Whole Earth Network and the Rise of Digital Utopianism*, that Stewart Brand, the editor of the *Whole Earth Catalogs* and *CoEvolution*

Quarterly, and many other publishing projects ever since, was one of the central actors in the emergence of what he calls digital utopianism. In the *Whole Earth Catalog* and the broader networks that Brand was involved with in the San Francisco bay area in the 1960s, computer liberation was mixed with Native American spirituality and ecological holism to contribute not just to technological development but to a cultural transformation, as well: the making of what Turner calls "cyberculture."

Like many of their fellow hackers in the 1960s and 1970s, Wozniak and Gates, who went on to found Apple and Microsoft, two of the largest companies in the world today, were not particularly active politically, but they were definitely affected by the widespread spirit of liberation that was so central to the social and cultural movements of the time. In our terms, they combined scientific knowledge and technical skills with cultural awareness into a hybrid imagination. In particular, they tried to find ways to develop the technology that they knew and loved in ways that could enhance people's power over their lives. Of course, as we all know, they would soon be subjected to market forces and commercializing pressures beyond belief and the tendency to hubris that came with them (even though Wozniak did not stay long in business). At the outset of the computer revolution, however, it was definitely the case that there was something more than merely a desire for making money that was going on in the head of many a radical innovator. As with Google, many years later, there was also a kind of hybrid imagination at work. After all, Google still has the corporate slogan, "don't be evil" and is using its technology and its various business ventures to contribute actively to the reduction of carbon dioxide emissions and lower the carbon footprint of its operations [Auletta, K., 2009].

6.5   FROM BIG SCIENCE TO TECHNOSCIENCE

At the same time as the computer revolution was taking place, other developments in science and technology, especially in molecular biology and genetics, were bringing the worlds of scientific theory into more intimate contact with engineering and technological development. Genetic engineering and the other so-called biotechnologies were among the first "technosciences" that would emerge in the coming decades. By mixing previously separated fields of knowledge into new combinations, these fields challenged both the traditional identities of scientists and engineers but also the traditional ways in which they were educated.

The molecular biology of the 1950s and, in particular, the double helix model of DNA constructed by James Watson and Frances Crick had initiated the process. The scientific-minded could use the so-called genetic code as a starting point for exploring the connections between particular kinds of genetic traits and particular kinds of diseases and plants, while the technically-minded could try to build apparatus that could transfer genetic material from one organism to another. There was a clear potential relevance both for agriculture and medicine – and many of the active scientists had been motivated in their research to develop higher-yield crops, as the so-called "green revolution" in India had done earlier. But if there was a social concern among many biotechnologists, there would soon be commercial opportunities that would be hard for others to resist.

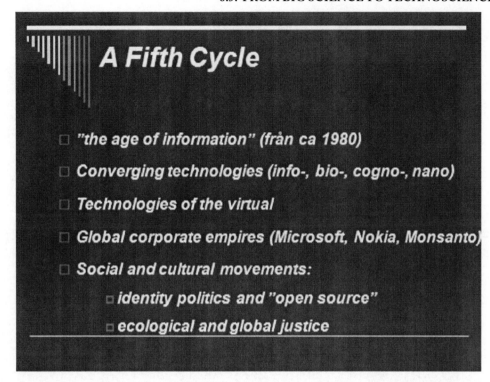

Already by the late 1970s, as it became clear to many observers that biotechnology had enormous economic potential, some of the scientists involved began to establish companies, where they could try to develop commercially viable products [Yoxen, E., 1983]. When Margaret Thatcher and Ronald Reagan were elected prime minister of Britain and president of the United States, this kind of "privatization" and entrepreneurial activity received government encouragement, as did other kinds of efforts to strengthen the interaction between universities and industries. In the 1990s, the international diffusion of the Internet, cellular telephones and other information and communications technologies contributed to an intensification of contacts between universities and technology firms, as well as to increasing attention to entrepreneurship and other aspects of knowledge management and product development.

Genetic engineering and information technology, and, most recently, nanotechnology, require expertise and skills from a number of scientific fields, as well as an engineering competence, put together in what might be termed a commercializable cocktail. While certainly not all science and technology has come to be integrated into processes of commercial innovation, there can be no denying that the rise of information technology and biotechnology industries has exerted a major influence on science and technology as a whole. As is readily apparent, these technosciences distinguish themselves from previous types of science and technology in at least three major respects (cf. Jamison, A. and Hård, M., 2003).

From Big Science to Technoscience

o *change in range and scope*

o *market orientation, corporate control*

o *university-industry collaboration*

o *entrepreneurial norm, or value system*

o *the state as strategist: innovation policy*

o *from assessment to promotion: "foresight"*

On the one hand, they are largely instrument-based technologies, which means that they require major expenditures on scientific research and, most especially, expensive scientific instruments – and computer power – for their eventual development. And unlike the science-based innovations of the early 20th century, which were, for the most part, applications of a scientific understanding of a particular aspect of nature (microbes, molecules, organisms, etc.), these new technologies are based on what Simon, H. [1969] once called the sciences of the artificial. Information technology is based on scientific understanding of man-made computing machines, and biotechnology is based on scientific understanding of humanly modified organisms. Nanotechnology is the most recent example of a "mode 2" field that was based on the development of scientific instruments to make a previously unreachable realm of reality available for commercial product development. The knowledge that is produced or constructed in these fields is thus about a simulated or technically mediated reality rather than the "natural" reality that existed in previous centuries.

Secondly, these are technologies that are generic in scope, which means that they have a wide range of potential applications in a number of different economic areas, social sectors and cultural life-worlds. As opposed to earlier generic technologies, or radical innovations – the steam engine, electricity and atomic energy, for example, which were primarily attempts to find solutions to identified problems - these new types of technologies tend to be solutions in search of problems. In this respect, information technologies, biotechnologies, and nanotechnologies are idea-based, rather than need-based, which means that, in relation to their societal uses, they are supply-driven, rather then demand-driven. That is one of the reasons why they require such large amounts of marketing and market research for their effective commercialization, and indeed for their development.

Finally, these advanced, or "high" technologies are transdisciplinary in what might be called their underlying knowledge base; that is, their successful transformation into marketable commodities requires knowledge and skills from a variety of different specialist fields of science and engineering. In earlier periods of technological development, there were clearer lines of demarcation between the specific types of competence and knowledge that were relevant; indeed, the classical categories of engineering are based on the particular types of scientific and technological theories that were utilized (chemical, mechanical, combustion, aerodynamic, etc.). Genetic engineering and information technology, and, most recently, nanotechnology, require expertise and skills from a number of scientific fields, as well as an engineering competence. The genetic engineer and the nanotechnologist certainly must know physics and chemistry and biology, but they do not know and learn these subjects in the same way as physicists, chemists and biologists. Rather they are taught to know what they need to know, in order to provide the society with new sorts of products.

In recent decades, many scientists and engineers have come to behave like businessmen, seeking out venture capital and market niches, and seeing each other as competitors rather than colleagues. These developments have led to a weakening – and, in many places, a complete elimination – of the sense of "enlightening" their fellow citizens that many makers of knowledge have traditionally thought they were doing. The idea, or, perhaps more accurately, the myth of a pure, or basic science, driven by curiosity and the "internal" motivations of the scientific practitioners themselves with an intrinsically beneficial societal function, has certainly not disappeared as an ideal. It has, however, become increasingly difficult to practice when the very possibility of carrying out scientific research has become dependent on the ever present need of funding and salesmanship. Not only have the borders between science and business become increasingly blurred, and the meanings and practices of science increasingly commercialized. At the same time, the procedures of public accountability and the criteria of scientific legitimacy have changed character.

At the close of the first decade of a new millennium science and technology find themselves at a critical juncture. In a time of economic recession with many national governments heavily in debt and levels of unemployment the highest in many years, there is once again, as in the 1960s, a need for questioning the assumptions that have guided policy making, as well as scientific and technological knowledge making over the past few decades. In particular, it seems important to question the dominant discourse of commercialization, with its overarching emphasis on linking scientists and engineers ever more intimately with the business world. This has involved both an institutional restructuring of universities, as well as a reshaping of many scientific and engineering fields so that they are more amenable to the needs and values of the commercial marketplace. The widespread fostering of entrepreneurship among scientists and engineers has certainly led to an effusion of new gadgets and high-tech wizardry, but it can be questioned whether this orientation has gone too far.

As we have mentioned in the introduction, many scientists and engineers have responded to the new conditions by trying to return to how it was, or is imagined to have been, in earlier times, when universities were "autonomous" and the role of science-based technology in society and in the economy was much more limited. The appeal of traditional ways of practicing science and technology has become quite strong, but as in previous periods of change, it seems counterproductive to think that the future can be met by returning to the past. Instead, as in the 1960s, there is a need for fundamentally rethinking and reconstructing the relations between science, technology and society, not least in order to meet the challenges of climate change and developing our societies, economies and communities in more sustainable ways, to which we now turn our attention in the final chapter.

CHAPTER 7

The Greening of Science and Technology

Our species is really quite special. We've learned so much and come so far in such a short time. So far, in fact, that it's all too easy to fall into the trap of hubris – of thinking that we really understand this world and that we can fix any problem that might come up. That is a dangerous assumption…As a society and as a species we've become so used to science and technology that we've forgotten that these are just ways of understanding and manipulating the world. They do not solve problems on their own. And as many problems as they do help resolve, they also create them anew.

David Suzuki and Dave Robert Taylor, *The Big Picture* (2009)

greening ['grining] *n*(Life Sciences & Allied Applications / Environmental Science) the process of making or becoming more aware of environmental considerations.

The Collins English Dictionary (1991)

7.1   THE MAKING OF GREEN KNOWLEDGE

An ambition to develop a greener kind of science and technology emerged as part of the social and cultural movements of the 1960s and 1970s. The word greening itself probably first reached the attention of a wider public in the best-selling book, *The Greening of America*, by Charles Reich, which began its life, as *Silent Spring* had done, and as so many non-fiction books still do today, as an article in the *New Yorker* a few months after the first Earth Day in 1970. Reich was not specifically concerned with environmental issues, however. He coined the term as a way to grasp what the revolts of the times were all about. As he put it:

> The logic of the new generation's rebellion must be understood in light of the rise of the corporate state under which we live and the way in which the state dominates, exploits, and ultimately destroys both nature and man. Americans have lost control of the machinery of their society, and only new values and a new culture can restore control. At the heart of everything is what must be called a change of consciousness. This means a new way of living–almost a new man…Industrialism produced a new man, too–one adapted to the demands of the machine. In contrast, today's emerging consciousness seeks a new knowledge of what it means to be human, in order that the machine, having been built, may now be turned to human ends [Reich, C., 1970].

In the course of the 1970s, green as an adjective, and greening as a noun, came to be used, perhaps especially in Germany, in a somewhat more specific way to refer to the new kind of political movements – and eventually political parties – that sprang up to protest against the deterioration of the natural environment in general and the risks and dangers of nuclear energy, in particular. And as the green movement spread around the world, there were some who started to envision what we will be characterizing in this chapter as the greening of science and technology.

In addition to the environmental and energy movements that became important political actors in many countries in the late 1970s, new departments for environmental studies, human ecology and environmental science were created throughout the world, and environmental research and technological development, particularly in relation to renewable energy, was given greatly increased attention and funding. Agencies for environmental protection and energy production were established at both the local, national and international levels, especially after the UN Conference on the Human Environment that was held in Stockholm in 1972. There was also within the cognitive praxis, or knowledge-making activities of the environmental movements a collective, or grass-roots kind of science and engineering, especially in regard to renewable energy and organic agriculture [Jamison, A., 2001, Jamison et al., 1990].

In this chapter, we trace the making of green knowledge through four phases: an initial period of awakening in the 1950s and 1960s, followed by politicization in the 1970s, particularly concerning energy politics. The 1980s and 1990s can be characterized as a phase of "normalization," as the forces of hubris and habitus led to a commercialization of green knowledge, on the one hand, and a professionalization, on the other. Most recently there has been a period of contention, particularly over climate change knowledge (cf. Jamison, A., 2010b).

7.2   A MIXING OF TRADITIONS

The environmental movements that emerged in the 1960s and 1970s represented a kind of collective hybrid imagination. They combined, according to the influential account by the historian Donald Worster in his book *Nature's Economy*, two opposing approaches to the environment, with funda-

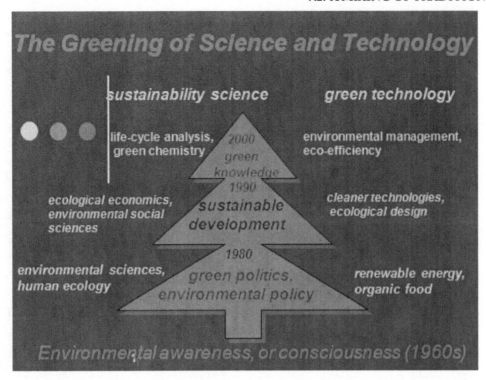

mentally different attitudes to nature, which he termed "imperialist" and "arcadian"[Worster, D., 1979]. Worster traced the scientific, or imperialist tradition back to the ideas of Francis Bacon in the early 17th century about the human domination of nature, which we discussed in chapter three. This human-centered, or anthropocentric, approach to the environment was developed further in the 18th century perhaps most influentially by Carl von Linné, or Carolus Linnaeus in Sweden in his system of classification that he imposed on the natural, or non-human world. In the course of industrialization, this highly utilitarian view of nature became the dominant approach to ecology and the environment, as science and engineering took on more professional and disciplined organizational form, and ecology itself became a science.

In his book, Worster pitted the imperialist tradition against what he called an "arcadian" tradition (the term was chosen in order to associate this alternative stream of ecological thought with the classical ideal of harmony between nature and society that Roman poets placed in the Greek region of Arcady). Worster traced the arcadian tradition back to the Romantic movement of the late 18th and early 19th centuries, which we discussed in chapter four, starting his book with the story of the English pastor and writer Gilbert White, and in particular White's *The Natural History of Selborne*, originally published in 1789. Worster then describes this alternative ecological tradition, primarily by focusing on the notebooks and natural history writings of Henry David Thoreau in

● ● ● | *A Brief History of Green Knowledge Making*

Awakening: 1960s
   Public education, criticizing (big) science

Politicization: 1970s
   Social movements, appropriate technology

Normalization: 1980s-90s
   Sustainable development, green business

Contention: 2000s-
   Dealing with climate change – and the skeptics!

the mid-19th century. He also discusses how in Germany and the Scandinavian countries, a related tradition of *Naturphilosophie* (or philosophy of nature) emerged in the late 18th century among academics and artists, and he shows that this romantic tradition had an influence in both geology and geography, biology and chemistry, and even in physics, where, as we have seen in chapter four, the search for an underlying "spirit" in nature led the Danish scientist Hans Christian Ørsted to look for and eventually discover the connection between the natural forces of electricity and magnetism.

Worster's argument was that the two ecological traditions, had met in Charles Darwin's theory of natural evolution, but that they had subsequently split apart, in the course of the 20th century, and developed into two different approaches to ecological science and environmental research. The one was systemic, leading to a large-scale, ecosystems oriented ecology, while the other was individual in focus, and had led to a smaller-scale, population-oriented ecology. The two traditions drew on different attitudes, or conceptions of nature, as well as different methodological and theoretical assumptions about how to investigate the natural environment and its non-human inhabitants.

It was these two different environmental, or ecological traditions, that were combined in the environmental movements of the 1960s and 1970s, according to Worster. But there was also a third tradition that was drawn upon, as well, namely the various "human ecologies" that had emerged in the 19th and early 20th centuries, both in Europe and in the United States. As we mentioned in chapter five, human ecology was an outgrowth of industrialization and urbanization and entered into the scientific and technological fields of geography and sociology, anthropology and planning, public health and civil engineering with the writings of George Marsh playing a particularly important role in the 19th century and those of Lewis Mumford having a formative influence in the 20th century. To understand the hybrid imaginations that were formed in the environmental movements of the 1960s and 1970s, it can therefore be useful to add this third tradition to Worster's two, and to distinguish three ecological traditions that were mobilized in the making of the environmental movements of the 1970s. Each tradition – the imperialist, the arcadian, and the human – has its own characteristic conception of nature and its own preferred forms of knowledge making.

Ecological Traditions			
	Imperialist	*Arcadian*	*Human*
Formative Influences	Francis Bacon Carl von Linné	Gilbert White Henry D Thoreau	George Marsh Lewis Mumford
Key Mobilizers	Odum brothers	Rachel Carson	Barry Commoner
Type of knowledge	systemic models managerial	natural history romantic	hybrid planning
Relation to nature	exploitation	participation	co-construction
Conception of nature	ecosystem resource base	community locality	region landscape

It would be these three traditions that were mobilized in the making of green knowledge in the 1960s and 1970s [Jamison, A., 2001]. On the one hand, the imperialist tradition was reinvented in the cybernetic language of ecosystems ecology and energy systems analysis. Ecosystem ecology, as developed by Eugene and Howard Odum, became extremely influential among natural scientists, particularly during the International Biological Program, and as a new approach to ecology, it would play a major role in the emergence of an environmental consciousness in the 1960s [Hagen, J., 1992].

In the World Wildlife Fund, founded in 1961, and then more scientifically in the International Biological Program in the mid-1960s, the "imperialist" tradition took on a more modern or contemporary manifestation. It became more explicitly international, as scientists and other conservationists came to take part in transnational research and development networks, particularly within the IBP project. The imperialist tradition was also brought up to date technologically with the new cybernetic and computer-based approaches to research that were developed by the new breed of ecosystem ecologists and were also utilized in the forecasts of resource and energy use that formed the basis of the book, *Limits to Growth*, that was published in 1972 by the Club of Rome.

It would be the biologist turned science writer Rachel Carson, whose eloquent writings would do most to give the arcadian tradition a contemporary resonance. "Over increasingly large areas of the United States," she wrote, "spring now comes unheralded by the return of the birds, and the early mornings are strangely silent where once they were filled with the beauty of bird song" (Carson 1962: 97). As we mentioned in the previous chapter, her book *Silent Spring* served to awaken the industrial world from its postwar slumbers, and she was soon followed by other writers who, with their scientific pondus and more sober tone, helped to reinvent the alternative ecological tradition movement and bring new life into the conservation organizations that had been established in the late 19th and early 20th centuries.

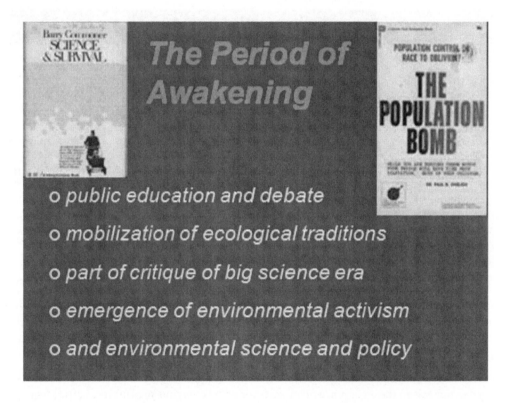

The mobilization of human ecology came from many different directions. Some, like Murray Bookchin, who would develop the concept of social ecology and write a number of influential books and articles in the following decades, brought a socialist sensibility into the environmental movement. His book from 1962, *Our Synthetic Environment*, which actually came out some six months before *Silent Spring*, was one of the first to present the entire range of new environmental problems, with particular emphasis on their social aspects; Bookchin devoted special attention to such issues as occupational health and safety, chemical pollution, household risks, waste disposal, etc.

Others, like the biochemist Barry Commoner, gave the environmental movement a more explicit focus on the role that science and technology played in the environmental crisis. Commoner depicted, in his first book, *Science and Survival* (1966), how science had come to cause as many, if not more problems than it solved, and suggested a number of public service, or critical, activities for scientists to play in the emerging movement. Later, in the 1970s, he would be one of the leaders in the anti-nuclear movement and would run for president as a candidate for the Citizens Party in 1980. Now in his nineties (he was born in 1917), Barry Commoner has served throughout his life as a role model for environmental "movement intellectuals" and an example of how science can be combined with cultural awareness [Egan, M., 2007].

Also important was the biologist Paul Ehrlich, who was one of the more active public educators in the period of awakening in the 1960s, with his best-selling book, *The Population Bomb* (1968). The different perspectives of Commoner and Ehrlich, the one emphasizing population growth while the other focused on technological and economic issues, would subsequently be combined in the range of new activist organizations that emerged in the 1970s, as well as in the new departments of environmental studies and environmental science that were established at many universities around the world.

By the end of the 1960s, ecology had inspired both the emergence of new activist groups, such as Friends of the Earth, as well as a process of policy reform and institution building. In the early 1970s, most of the industrialized countries established new state agencies to deal with environmental protection, as well as environmental research and technological development. Many national parliaments also enacted more comprehensive environmental legislation and, at the United Nations Conference on the Human Environment in Stockholm in 1972, protecting the environment was recognized as a new area of international concern.

A biologist, René Dubos, and an economist, Barbara Ward, collaborated on the book that would set the agenda for the conference, *Only One Earth*. It was, in its own way, an example of hybridity, as the new kind of environmentalism that Ward and Dubos proposed combined a call for efficient management of resources with an empathetic understanding of life and nature: "Now that mankind is in the process of completing the colonization of the planet," they wrote, "learning to manage it intelligently is an urgent imperative. Man must accept responsibility for the stewardship of the earth" [Ward, B. and Dubos, R., 1972, p. 25]. They admitted that the proposals that they suggested in their book would not come easily: "the planet is not yet a centre of rational loyalty for all mankind. But possibly it is precisely this shift of loyalty that a profound and deepening sense of our shared and inter-dependent biosphere can stir to life in us" (Ibid: 298).

7.3   A PERIOD OF POLITICIZATION

It was the so-called oil crisis, of 1973-74, when the OPEC countries raised substantially the price of oil that triggered a major shift in green knowledge making, as energy issues moved to the top of many national political agendas. Particularly in North America and western Europe, the 1970s marked a decade of intense political debate and social movement activity, as the pros and cons of different energy options, in particular nuclear energy, were contested in public. Seen in retrospect, an important result of the energy debates of the 1970s was a professionalization of environmental concern and an incorporation by the established political cultures of what had originally been a somewhat delimited political issue. For a brief period in the late 1970s, the environmental movement, especially in Scandinavia, Germany and the Netherlands, became a social and political force of major significance [Jamison et al., 1990]. When the issues that inspired the movement were resolved, and taken off the political agenda, the different component parts tended to split apart and fragment. The unity that had been achieved in struggle could simply not be sustained.

The cognitive praxis of the environmental movements was based on a philosophy, or cosmology, of systemic holism derived from systems theory and popularized in such books as Barry Commoner's *The Closing Circle* (1971) and *Only One Earth* [Ward, B. and Dubos, R., 1972], as well as in *A Blueprint for Survival* (The Ecologists 1972), which launched the journal, *The Ecologist*, and the extremely influential *Limits to Growth*. Barry Commoner's four laws of ecology – "everything is connected to everything else," "everything must go somewhere," "nature knows best," and "there is no such thing as a free lunch" – provided a set of cosmological, or world-view assumptions, for the environmental movements that, in the course of the 1970s, became significant political actors

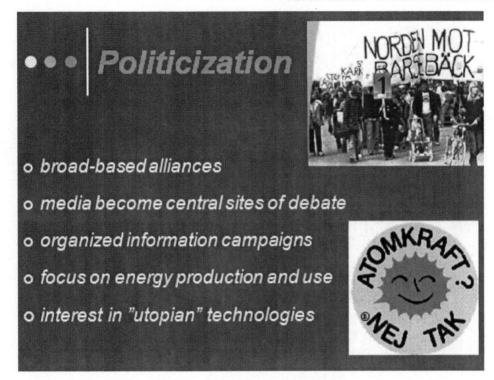

in several northwestern European countries, as well as in North America. In political campaigns directed against various kinds of air and water pollution, chemicals in food and agriculture, and especially against the development of nuclear energy, environmental movement organizations, together with students and teachers at universities, began to turn scientific knowledge and technological development green.

In the environmental movements of the 1970s, an ecological philosophy, or world-view, was combined with a practical interest in appropriate, small-scale technology that was popularized in such books as *Small is Beautiful* by E.F. Schumacher, *Tools for Conviviality* by Ivan Illich and *Alternative Technology and the Politics of Technical Change*, by David Dickson and practiced in new movement settings, or public spaces, such as the Center for Alternative Technology in Wales, the New Alchemy Institute in the United States, and a wide range of production collectives and alternative communities, as we briefly mentioned in the previous chapter. At these sites, environmental and energy activists could learn about "environmentally-friendly" ways to produce energy, food, and the other necessities of life that were based on an ecological worldview. Activists and academics joined together to build solar energy panels and wind energy plants, grow organic food, and try to live more ecologically [Boyle, G. and Harper, P., 1976]. In the Netherlands, "science shops" were established at several universities to provide meeting places between the academic world and the

The Cognitive Praxis of Environmental Movements

- ● ● ●

  o *Cosmological dimension:*
     systemic holism, "limits to growth"

  o *Technological dimension:*
     appropriateness, "small is beautiful"

  o *Organizational dimension:*
     participatory research, "citizen science"

broader society, and in many other countries, the environmental movements fostered other forms of what the sociologist Alan Irwin later termed "citizen science"[Irwin, A., 1995].

A kind of "grass-roots" engineering emerged in many parts of the world, particularly in relation to the anti-nuclear energy movements. In Denmark, scientists and engineers created a national Organization for Renewable Energy (or OVE, *Organisation for vedvarerende energi*) that helped people throughout the country to learn how to build their own wind energy plants and solar panels [Jamison, A., 1978]. OVE arranged courses at older as well as newly-established folk high schools, and created centers for renewable energy, such as the Nordic center in Thisted, which is still in operation.

In 1978, the world's then largest wind energy power plant was constructed by students at the Tvind folk high schools on the Danish west coast, not far from where VESTAS is now based. Mobilizing a Danish tradition – as we mentioned in chapter four, Poul La Cour, a folk high school physics teacher in the 19th century had been one of the first in the world to experiment systematically with wind-power generated electricity production – the Organization for Renewable Energy has continued to foster "grass-roots innovation" ever since. By the late 1970s, the movement had spawned a number of companies, one of which, VESTAS, is now the leading wind turbine producer in the world and one of Denmark's largest companies.

Appropriate Technology in the 1970s

The New Alchemy Institute Ark

Nordic Folkcenter for Renewable Energy

7.4   NORMALIZATION AND THE RISE OF GREEN BUSINESS

In the 1980s, as the political climate in North America and northwestern Europe turned to the right, environmental politics changed character, and the making of green knowledge changed as well. From a cultural perspective, this right turn in politics represented a mobilization of conservative traditions, or – as they are often referred to in the United States – neo-conservative values and interests. Traditional religious and nationalist concerns were fundamental to these neo-conservative movements, which emerged, at least in part, as a kind of organized opposition to the environmental and women's movements of the 1970s and the kind of knowledge they had embodied and articulated [Helvarg, D., 1988, Rowell, A., 1996].

In many European countries, similar movements emerged at this time to oppose immigration and European integration. In Denmark, there was a strong mobilization against entrance into the European Union, and this later led to the building of the Danish People's Party which, in many ways, retains the character of a social movement even though it has become an established political party. Neo-nationalism in Europe resembles neo-conservatism in the United States, both in terms of an adherence to what might be termed a populist conception of knowledge, as well as in regard to a cosmological belief in national identity and the importance of upholding traditional values.

It was within the cultural space carved out by these neo-conservative and neo-nationalist movements that anti-environmentalism would emerge as a political force in the course of the 1980s. Already in the debates about nuclear energy in the 1970s, a number of natural scientists, especially atomic physicists, challenged the forms of knowledge-making that were promulgated in the new social movements of the 1970s; indeed, the energy debates of the 1970s were, in large measure, debates about different conceptions of science and technology.

At the same time as the anti-environmental "backlash" was taking shape in the 1980s, the environmental movement itself fragmented into a number of different organizations and institutions, both in terms of politics and knowledge-making. Green parties were formed in many countries and professional activist organizations, such as Greenpeace, grew in significance, while the broad-based, or grass-roots, organizations that had led the campaigns against nuclear energy in the 1970s tended to be marginalized. Within universities and new environmental "think tanks" such as the World Resources Institute and the Wuppertal Institute, environmental and energy experts started to make more specialized kinds of knowledge in renewable energy, organic agriculture, and eventually in relation to climate change [Jamison, A., 1996].

As such, more professional and established forms of knowledge-making started to replace the kinds of appropriate or alternative science and technology that had been so prominent in the 1970s. Many of those who had been active in the environmental and energy movements in the 1970s left the movement "space" behind to make careers in universities, as well as in the wider worlds of government, media, and business.

In 1987, the report, *Our Common Future*, was published by the World Commission on Environment and Development, headed by the former Norwegian prime minister, Gro Harlem Brundtland and with representatives from government, business, academia, as well as from environmental think-tanks and so-called non-governmental organizations. With its call for "sustainable development" – by which was meant a kind of socio-economic development that took into account the needs of future generations for natural resources - the report signaled the coming of a new international political doctrine in which environmental concern was to be included into all other areas of social and economic life.

Following the fall of the Soviet empire, and the so-called Earth Summit in Rio de Janeiro in 1992 (the UN Conference on Environment and Development), where the idea of sustainable development was translated into the Agenda 21 document, new approaches to greening science and technology proliferated in the 1990s. Particularly prominent have been the efforts to encourage what are often termed market-oriented solutions to environmental problems. The general approach can be thought of as an incorporation of environmental concern into the world of business. There has emerged a range of activities in such areas as environmental management, cleaner technology, industrial and urban ecology, and green product development, which are explicitly commercial. As a result, instead of learning together and cooperating with each other in projects of collective learning, as they did in the 1970s, many makers of green knowledge have gone into business during the past 20 years.

These forms of knowledge-making became especially important in several European countries, where social-democratic governments, often with the support of green parties, pursued policies of "ecological modernization" as did the Clinton-Gore administration in the United States. In Germany, Great Britain, Denmark, Sweden and the Netherlands, as well as at the European Commission, ecological modernization sought to combine environmental concern with economic growth. As climate change became a more integral part of environmental politics in the 1990s, it was the market-oriented approaches that tended to dominate the international deliberations, both in Kyoto, as well as within intergovernmental administrative and scientific advisory bodies, such as the Intergovernmental Panel on Climate Change (IPCC).

The rise of market-oriented environmentalism, or green business, was shaped by the broader neo-liberal movement, which has provided the dominant story-line of the past two decades, both in regard to science and technology, in general, and environmental science and technology, in particular [Hoffman, A., 2001]. Much of the knowledge-making activity within green business tends to be organized in commercial networks, with university scientists and engineers working together with companies on specific projects. There are also a number of "movement intellectuals" in the commercial media as well as in private consulting companies who serve to articulate the underlying importance of meeting the sustainability challenge in commercial terms. The "cognitive praxis" of

Dialectics of Sustainable Development

Green business	Critical ecology
"Ecological modernization"	"Environmental justice"
Instrumental rationality	Communicative rationality
Technological innovation	Appropriate technology
Commercial orientation	Community emphasis
Expert solutions	Public engagement

green business exemplifies the dominant approaches of academic capitalism in the promotion of commercially-oriented technological innovation and green product development.

The cosmology of green business is based on a belief in a convergence between economic growth and environmental protection, and depending on the context, it has been termed ecological modernization, eco-efficiency, corporate sustainability, or green growth. In the words of Maarten Hajer, what was central to the political discourse of ecological modernization in the 1990s was "the fundamental assumption that economic growth and the resolution of the ecological problems, can in principle, be reconciled. Hence, although some supporters may individually start from moral premises, ecological modernization basically follows a utilitarian logic: at the core of ecological modernization is the idea that pollution prevention pays" (Hajer 1995: 27). In the course of the past fifteen years, particularly in China and other Asian countries, this fundamental assumption is central to major national programs in "green growth."

In relation to climate change, one of the main proponents of market-oriented, or green business, approaches has been the former U.S. vice-president Al Gore. Already in his first book, *Earth in the Balance*, from 1993, written while he was still a Senator, Gore combined arguments for economic growth with arguments for environmental protection in providing what he called a "new common

Green Business as Cognitive Praxis

● ● ●

From "movement"...	to "institutions"
appropriate technology	green products
organizational alliances	competing networks
ecological society	sustainable growth
popular education	entrepreneurship
knowledge integration	specialization
movement intellectuals	commercial brokers

purpose" for humanity. After the fall of the Soviet empire, the "singular will of totalitarianism" had fallen as a challenge:

> But now a new challenge – the threat to the global environment – may wrest control of our destiny away from us. Our response to this challenge must become our new central organizing principle. The service of this principle is consistent in every way with democracy and free markets [Gore, A., 1993, p. 277].

In his book, Gore proposed what he then termed a "Global Marshall Plan" for saving the environment, by which he meant massive investments in renewable energy companies and in other environmentally-friendly technological developments. In the 1990s, as vice-President, Gore led the U.S. delegation to Kyoto, where he was one of the central promoters of what has since been termed the cap-and-trade approach for dealing with climate change. After losing the 2000 election, Gore emerged as the main proponent for using market mechanisms and business ventures to respond to what he so famously called the "inconvenient truth" of climate change.

7.5   COMBINING ENVIRONMENTALISM AND JUSTICE

Since the late 1990s, a new kind of political activism, often involving forms of civil disobedience and direct action, has emerged in relation to environmental issues and, most recently climate change, as a part of what has been characterized as a broader movement for global justice (cf. Jamison, A., 2010b).

The global justice movement has been called a "movement of movements," a term coined by Naomi Klein in the wake of the anti-globalization protests of the late 1990s, and captures well the heterogeneous character of the activists concerned with environmental and climate justice, as well as the broader global justice movement [Klein, N., 2000]. Both are filled with tensions and contradictions, composed as they are of a variety of groups and individuals who have begun to take political action in order to protest the quite different kinds of negative consequences that they attribute to globalization, and they propose ways of dealing with them in a more equitable, or just manner.

For the influential theorists Hardt, M. and Negri, A. [2004], the working class or "masses" that were mobilized in the social movements of the late 19th and early 20th centuries have given way to a "multitude" of disenfranchised and disenchanted global citizens. While a multitude of voices and concerns has begun to be heard in relation to globalization, the multitude has not yet formed a shared set of beliefs that can serve as a cosmological dimension for a social movement's cognitive praxis. Like other social movements in their initial stages, there is as yet no real integration of the relatively abstract theorizing about global justice voiced by theorists like Hardt and Negri with the multifaceted array of practical activities that are being carried out; there is not yet a social movement with a coherent or integrated cognitive praxis.

In addition to the political protests, which became most visible, in relation to sustainability issues, in the streets of Copenhagen at the end of 2009 during the COP15, there are a number of primarily local organizations in both the global North and global South that carry out a range of more constructive activities in relation to such areas as renewable energy, ecological housing and design and organic agriculture. In recent years, there have been attempts to arrange gatherings, where the different component parts of the global justice movement can meet and discuss their concerns, and exchange their experiences. These various "social forums," as they have come to be called, have taken place both at an international level (at world social forums, that have been held each year since 2000), as well as at more regional, national, and local levels, particularly in Europe [Fisher, W. and Ponniah, T., 2003].

There are geographical tensions among the various component parts of the emerging global justice movement, and there are major differences among those actively involved in regard to how to connect sustainability and climate change to the broader concerns with global justice. Sustainable, or climate justice, tends to mean something very different for activists in the global North than it does for activists in the global South. The very different life experiences and expectations of the participants make it difficult to develop a common understanding and shared belief system. Ideas of fairness and equity are highly dependent on contexts of history and place [Parks, B. and Roberts, J.T., 2010].

In addition to this basic geographical conflict, there are also generational and intellectual tensions. The institutionalized legacies of the new social movements of the 1970s – primarily the larger environmental NGOs (or, non-governmental organizations) – tend to see climate change and sustainability exclusively as an environmental challenge and, until quite recently, have tended to disregard the social and political implications of climate change. The more socially minded activists, on the other hand, often working in development assistance NGOs, have, on the other hand, tended not take environmental and, more recently, climate issues, all that seriously. There has been a sectorial division of labor that can be seen as an unfortunate consequence of the professionalization processes that affected the greening of science and technology in the 1980s and 1990s.

The task of alerting the public to the wide range of challenges that climate change and sustainable development more generally raise in regard to global justice and social inequality has fallen primarily to a relatively small group of newer organizations and activists. Particularly in Africa, Asia and Latin America, groups and alliances to save rainforests, preserve biodiversity, defend the rights of indigenous peoples, and develop sustainable forms of agriculture and industry have been developing since the 1970s, and in North America and some European countries, "environmental justice" groups have emerged in many minority communities and neighborhoods to combine environmental concern with struggles against racism and discrimination [Gottlieb, R., 2001, Schlosberg, D., 2007, Taylor, B., 1995].

There are also a growing, but still relatively small, number of cases of collaboration between academics and activists in universities and local communities in trying to deal with climate change and other environmental problems in just or equitable ways [Hess, D., 2007, Worldwatch Institute, 2010]. New forms of community-based innovation and knowledge-making can be identified in local food movements around the world, as well as in a range of not-for-profit engineering projects in such areas as sustainable transport, renewable energy, and low-cost, environmentally-friendly housing. Such projects as the Alley Flat Initiative at the University of Texas in which students and teachers from the School of Architecture have designed low-cost, climate-smart housing in East Austin in cooperation with local housing suppliers and neighborhood groups show what can be done [Jamison, A., 2009].

The Alley Flat Initiative emerged as part of a larger project on sustainable development, directed by architecture and planning professor Steven Moore. The ongoing project includes a design studio not only taught by Moore but also by Louise Harpman and a visiting professor, Sergio Palleroni, who had previously carried out community-oriented architectural projects with students at the University of Washington. Looking for a specific focus for the studio, the students spent time in East Austin, the area of the city that in the early 20th century had been segregated through the provisioning of infrastructure as an African-American, Latino and industrial area. Like many such areas in many American cities, east Austin became threatened by so-called gentrification when

Blacks and Latinos cleaned up industrial brownfields over six or seven decades, making the area attractive to more wealthy whites.

For example:
The Alley Flat Initiative

The Alley Flat Initiative is a joint collaboration between the University of Texas Center for Sustainable Development, the Guadalupe Neighborhood Development Corporation, and the Austin Community Design and Development Center. The Alley Flat Initiative proposes a new sustainable, green affordable housing alternative for Austin.

The motivation behind the initiative was to find a way to learn architecture by doing something useful for the community, and after looking through maps and reading about the history of the area the students came up with the idea of designing climate-smart alley flats, or second houses along the alleys – what used to be called "granny flats" because they were where grandparents lived – that could help the residents pay their escalating property taxes and fight off gentrification, and also contribute to the transition to a low-carbon society. As described on the initiative's website:

> The initial goal of the project was to build two prototype alley flats – one for each of two families in East Austin – that would showcase both the innovative design and environmental sustainability features of the alley flat designs. These prototypes were built to demonstrate how sustainable housing can support growing communities by being affordable and adaptable. The first of these prototypes celebrated its house warming with the community in June of 2008 and the second prototype was completed in August of 2009. The long-term objective of the Alley Flat Initiative is to create an adaptive and self-perpetuating delivery system for sustainable and affordable housing in Austin. The "delivery system" would include not only efficient housing designs constructed with

sustainable technologies, but also innovative methods of financing and home ownership that benefit all neighborhoods in Austin [AFI, 2011].

Moving into the second alley flat...

Unfortunately, however, such activities fall well outside of the mainstream and remain quite marginal at universities throughout the world, although, in recent years, several universities in the United States have established programs in engineering for sustainable community development [Lucena et al., 2010]. In some of these programs, there is a similar kind of institutional outreach that was so characteristic of the "movement" activities that took place in the 1970s, but most of them have yet to achieve the influence and legitimacy that would make them significant players in the global, or local, politics of climate change mitigation and sustainable development. The increasing encroachment of a commercial and entrepreneurial value system at universities makes it difficult for concerns with social justice to be given the attention they deserve in science and engineering education.

7.6   CONTENDING APPROACHES TO GREENING

The greening of science and technology can be seen as an ongoing process of contention between three approaches or strategies that correspond to what we have termed in chapter two as the contending cultures of science and engineering. The dominant approach to greening can be considered a part of what has been termed the new "mode" of knowledge production, or "mode 2" in which the borders between the academic and business worlds are increasingly transgressed [Gibbons et al., 1994]. In these contexts, science is not carried out in a disinterested and impartial fashion, but it is rather funded by external interests in order to contribute directly to commercial innovations. Such research is often carried out in networks connecting academics and companies in specific projects, in order to develop profitable green products and provide commercial "solutions" to climate change and other environmental problems.

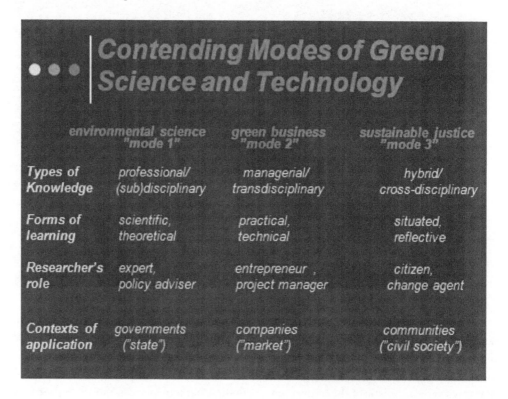

	environmental science "mode 1"	green business "mode 2"	sustainable justice "mode 3"
Types of Knowledge	professional/ (sub)disciplinary	managerial/ transdisciplinary	hybrid/ cross-disciplinary
Forms of learning	scientific, theoretical	practical, technical	situated, reflective
Researcher's role	expert, policy adviser	entrepreneur , project manager	citizen, change agent
Contexts of application	governments ("state")	companies ("market")	communities ("civil society")

On the other hand, there is an academic or professional approach to greening knowledge in the name of sustainability, or environmental science, that is based on a more traditional conception of objective and impartial expert knowledge. In this culture of science and engineering, greening tends to be subdivided into the more traditional scientific disciplines and engineering fields, into sub-disciplines, and the knowledge that is produced tends to be highly specialized, according to certain methods or paradigms. This is often a form of what Thomas Kuhn referred to in the 1960s as "normal science."

A third approach, or mode 3, that explicitly connects the quest for sustainable development to concerns of global justice and fairness in what I have termed "change-oriented research" is comparatively weak at the present time [Jamison, A., 2010a]. While a number of good examples have provided sources of inspiration and mobilization in recent years, perhaps especially in relation to ecological, or organic food and sustainable, or renewable energy, there has not yet developed a commonly shared theoretical and conceptual framework for such action, or change-oriented sustainability research.

Since climate change and sustainability in general are such all-encompassing and multifaceted issues, it will be necessary in this emerging third mode of greening science and technology to foster what we have termed in this book a "hybrid imagination," mixing natural and social, local and global, academic and activist forms of knowledge in new combinations. In this regard, the efforts of the physicists-turned-environmentalists, Vandana Shiva and Fritjof Capra, can provide inspiring role models.

Dr. Shiva speaking at a rally at Monsanto's headquarters in St. Louis. Photo by Nic Paget-Clarke.

Vandana Shiva and Carlo Petrini at a Slow Food Cafe

In their writings, both Shiva and Capra have tried through many years to combine disparate fields of science and technology in an engaged and personal way, and they have presented their knowledge in a popular, accessible form, outside of the established academic world (e.g., Capra, F., 1982, 2002, 2007). Shiva has written about science, technology and development [Shiva, V., 1988], and more recently has written about the risks involved, both to nature and humanity, in genetically-modified foods [Shiva, V., 2000]. She has also coined the term, "earth democracy" as a way to characterize the broader ambition to combine environmentalism, or sustainable development, with the quest for global justice [Shiva, V., 2005].

Fritjof Capra left a career in physics in the 1970s to write his extremely influential book on the connections between eastern religion and modern physics (*The Tao of Physics*) and then has ever since taken active part in the environmental movements in both Europe and the United States, where he has settled. He has written a series of popular science books on environmental issues and on what he has termed the "hidden connections" between processes in nature and society, and most recently, he has published a book on the science of Leonardo da Vinci, presenting the unpublished notebooks that Leonardo left behind.

Both Shiva and Capra have also created centers for research and education and taken part in a wide range of political campaigns and struggles, in Shiva's case, especially in relation to food and agriculture and, more recently, nanotechnology and climate change, and in Capra's case, in relation to what he has termed sustainable living. At the Center for Ecoliteracy in Berkeley, Capra and his collaborators have developed teaching material and educational programs for schools and local communities to help cultivate more ecological, or sustainable, life-styles and forms of behavior.

In the future, it can only be hoped that the kinds of cognitive praxis that are beginning to emerge in the climate and sustainable justice movements, building on the examples of Vandana Shiva, Fritiof Capra – and others such as David Suzuki, who has been a pioneer in public interest science journalism and television broadcasting in Canada – will be able to grow and, not least, achieve much greater legitimacy and acceptance among scientists and engineers throughout the world.

A Hybrid Imagination:
David Suzuki
biologist-broadcaster
academic- activist

David Suzuki

the BIG
PICTURE

From the website of "The Nature of Things with David Suzuki",
the television program that he has hosted since 1979 in Canada

Much will depend on the ways in which science and engineering education and universities more generally are made more "sustainable." There are many efforts around the world, but many of them are more rhetorical than real, more concerned with branding and image-building and what has been called "green-washing" than with substantive integration of contextual knowledge into educational programs. There needs to be more room, or space, at universities for students and teachers to undertake "not-for-profit," community-oriented activities in addition to the more traditionally academic and the increasingly dominant commercial activities that are carried out in both research and education. In a world in which universities have become ever more subjected to the forces of hubris in the guise of "global competitiveness," cross-disciplinary and cross-cultural knowledge making is, to put it mildly, not particularly encouraged, well-supported, or understood. If scientists and engineers are to meet the challenges that they face in a spirit of critical engagement, it will be crucially important in the years to come to see to it that our universities can help to foster hybrid imaginations among science and engineering students.

Bibliography

AFI (2011). The website of the Alley Flat Initiative, http://www.thealleyflatinitiative.org/accessedMarch29,2011. 145

Auletta, Ken (2009). *Googled. The End of the World as We Know It*. London: Penguin Books. 122

Baillie, Caroline (2009). *Engineering and Society: Working Towards Social Justice. Part 1. Engineering and Society*. Morgan & Claypool. 1

Bhabha, Homi (1994). *The Location of Culture*. London: Routledge. 2

Bijker, Wiebe (1995). *Of Bicycles, Bakelites and Bulbs. Toward a Theory of Sociotechnical Change*. Cambridge, Ma: The MIT Press. 22

Bijker, Wiebe, Thomas P. Hughes and Trevor Pinch, eds., (1987). *The Social Construction of Technological Systems. New Directions in the Sociology and History of Technology*. Cambridge, Ma: The MIT Press. 18

Bok, Derek (2003). *Universities in the Marketplace: The Commercialization of Higher Education*. Princeton University Press. 8

Borish, Steven (1991). *The Land to the Living : The Danish Folk High Schools and Denmark's Non-violent Path to Modernization*. Nevada City, CA: Blue Dolphin Publishing. 65

Bourdieu, Pierre (2004). *Science of Science and Reflexivity*, translated by Richard Nice. Cambridge: Polity. 8

Boyle, Godfrey and Peter Harper, eds., (1976). *Radical Technology*. Wildwood. 135

Brown, Alison (1999). *The Renaissance (second edition)*. London: Longman. 27, 35

Bush, Vannevar (1945/1960). *Science, the Endless Frontier: A Report to the President for Postwar Scientific Research*. The National Science Foundation. 111

Castells, Manuel (1996). *The Rise of the Network Society*. Blackwell. 107

Capra, Fritjof (1982). *The Turning Point: Science, Society and the Rising Culture*. Simon and Schuster. 147

Capra, Fritjof (2002). *The Hidden Connections*. Doubleday. 147

Capra, Fritjof (2007). *The Science of Leonardo*. New York: Anchor Books. 39, 147

Christensen, Steen Hyldgaard and Erik Ernø-Kjølhede (2006). Reengineering Engineers, in Jens Christensen, Lars Bo Henriksen and Anette Kolmos, eds., *Engineering Science, Skills and Bildung*. Aalborg University Press. 10

Christianson, John (2000). *On Tycho's Island: Tycho Brahe and His Assistants, 1570–1601*. Cambridge University Press. 41

Cipolla, Carlo (1965). *Guns, Sails and Empires. Technological Innovation and the Early Phases of European Expansion, 1400–1700*. New York: Pantheon. 29

Cumming, Elizabeth and Wendy Kaplan (2002). *The Arts and Crafts Movement*. London: Thames and Hudson. 73

Dalton, Dennis (1993). *Mahatma Gandhi. Nonviolent Power in Action*. New York: Columbia University Press. 96, 99

della Mirandola, Giovanni Pico (1486). Oration on the dignity of man. `www.cscs.umich.edu/~crshalizi/Mirandola`. Accessed February 10, 2011. 38

Dickens, Charles (2000/1854). *Hard Times*. London: Wordsworth Classics. 59

Dickson, David (1974). *Alternative Technology and the Politics of Technical Change*. Glasgow: Fontana. 24

Dutta, Krishna and Andrew Robinson (1996). *Rabindranath Tagore: The Myriad-Minded Man*. New York: St. Martin's. 97

Downey, Gary (2010). *What is Global Engineering Education For? The Making of International Educators, Part 1*. Morgan & Claypool. 1

Egan, Michael (2007). *Barry Commoner and the Science of Survival*. Cambridge, Ma: The MIT Press. 133

Eisenstein, Elizabeth (1983). *The Printing Revolution in Early Modern Europe*. Cambridge University Press. 29

Elzinga, Aant (1985). Research, bureaucracy and the drift of epistemic criteria, in Björn Wittrock and Aant Elzinga, eds., *The University Research System. The Public Policies of the Home of Scientists*. Stockholm: Almqvist & Wiksell. 8

Elzinga, Aant (2002). The Growth of Science. Romantic and Technocratic Images. Accessed March 28, 2011 at `http://www.autodidactproject.org/other/aant3.html` 61

Elzinga, Aant and Andrew Jamison (1984). Making Dreams Come True: An Essay on the Role of Practical Utopias in Science, in Everett Mendelsohn and Helga Nowotny, eds., *Nineteen Eighty-Four: Science between Utopia and Dystopia*. Dordrecht: Reidel. 43

EU Nanotechnology (2011). `http://cordis.europa.eu/nanotechnology/` accessed March 29, 2011. 107

Eyerman, Ron and Andrew Jamison (1991). *Social Movements. A Cognitive Approach*. Cambridge: Polity. 24

Eyerman, Ron and Andrew Jamison (1998). *Music and Social Movements*. Cambridge University Press. 24

Fisher, William and Thomas Ponniah, eds., (2003). *Another World is Possible. Popular Alternatives to Globalization at the World Social Forum*. London: Zed Books. 142

Flyvbjerg, Bent (2001). *Making Social Science Matter. Why social inquiry fails and how it can succeed again*. Cambridge University Press. 5

Freeman, Christopher (1974). *The Economics of Industrial Innovation*. Harmondsworth: Penguin. 19, 113

Freeman, Christopher (1987). *Technology Policy and Economic Performance. Lessons from Japan*. London: Pinter. 19, 53

Freeman, Christopher and Francisco Louçá (2003). *As Time Goes By. From the Industrial Revolutions to the Information Revolutions*. Oxford University Press. 19, 106

Galison, Peter (2003). *Einstein's Clocks, Poincaré's Maps*. London: Sceptre. 81

Gandhi, M. K. (1927). *An Autobiography, or The story of my experiments with truth*. Ahmedabad: Navajivan Publishing House. 96

Gay, Peter (2009). *Modernism. The Lure of Heresy from Baudelaire to Beckett and Beyond*. London: Vintage Books. 86, 90

Gibbons, Michael, Camille Limoges, Helga Nowotny, Simon Schwartzman, Peter Scott, and Martin Trow (1994). *The New Production of Knowledge. The Dynamics of Science and Research in Contemporary Societies*. London: Sage. 2, 3, 104, 146

Giddens, A (2009). *The Politics of Climate Change*. Polity.

Gimpel, Jean (1976). *The Medieval Machine. The Industrial Revolution of the Middle Ages*. Harmondsworth: Penguin. 29

Gore, Al (1993). *Earth in the Balance*. Plume. 141

Gottlieb, Robert (2001). *Environmentalism Unbound. Exploring New Pathways for Change.* Cambridge, Ma: The MIT Press. 143

Grafton, Anthony (2000). *Leon Battista Alberti: Master Builder of the Italian Renaissance.* New York: Hill and Wang. 34

Grafton, Anthony and Ann Blair, eds., (1990). *The Transmission of Culture in Early Modern Europe.* Philadelphia: University of Pennsylvania Press. 39

Greisman, H. C. (1976). Disenchantment of the world: Romanticism, aesthetics and sociological theory, in *The British Journal of Sociology.* Vol. 27, No. 4. DOI: 10.2307/590188 61

Grell, Ole Peter, ed. (1998). *Paracelsus: The Man and His Reputation, His Ideas and Their Transformation.* Leiden: Brill. 39

Gropius, Walter (1918). *Bauhaus – Manifest.* Weimar. Flugblatt. 90

Hagen, Joel (1992). *An Entangled Bank. The Origins of Ecosystem Ecology.* New Brunswick, NJ: Rutgers University Press. 131

Hammond, Kim (2004). Monsters of modernity: Frankenstein and modern environmentalism, in *Cultural Geographies.* Vol. 11, No. 2. DOI: 10.1191/14744744004eu301oa 59

Haraway, Donna (1991). A Cyborg Manifesto: Science, Technology and Socialist-Feminism in the Late Twentieth Century, in *Simians, Cyborgs and Women: The Reinvention of Nature.* New York: Routledge. 2

Hård, Mikael and Andrew Jamison (2005). *Hubris and Hybrids: A Cultural History of Technology and Science.* New York: Routledge. 3, 9

Hardt, Michael and Antonio Negri (2004). *Multitude.* Penguin. 142

Headrick, Daniel (1981). *The Tools of Empire. Technology and European Imperialism in the Nineteenth Century.* Oxford University Press. 29

Helvarg, David (1988). *The War Against the Greens.* Sierra Club Books. 137

Herf, Jeffrey (1984). *Reactionary Modernism: Technology, Culture, and Politics in Weimar and the Third Reich.* Cambridge University Press. 84

Hess, David (2007). *Alternative Pathways in Science and Industry. Activism, Innovation and the Environment in an Era of Globalization.* The MIT Press. 143

Hill, Christopher (1965). *Intellectual Origins of the English Revolution.* Oxford: Clarendon. 43

Hill, Christopher (1975). *The World Turned Upside Down: Radical Ideas during the English Revolution.* Harmondsworth: Penguin. 44

HLEG (2004). Converging Technologies – Shaping the Future of European Societies. A report written by a working group, Foresighting the Next Technology Wave (Alfred Nordmann, *rapporteur*). Brussels: European Commission. 106

Hobsbawm, Eric (1962). *The Age of Revolution. Europe 1789–1848*. London: Weidenfeld & Nicholson Ltd. 56

Hobsbawm, Eric (1975). *The Age of Capital 1848–1875*. London: Weidenfeld & Nicholson Ltd. 66

Hobsbawm, Eric (1987). *The Age of Empire 1875–1914*. London: Weidenfeld & Nicholson Ltd.

Hofman, Andrew (2001). *From Heresy to Dogma: An Institutional History of Corporate Environmentalism*. Palo Alto, CA: Stanford University Press. 139

Holmes, Richard (2008). *The Age of Wonder. How the Romantic Generation Discovered the Beauty and Terror of Science*. London: HarperPress. 62

Hounshell, David (1984). *From the American System to Mass Production*. Baltimore: Johns Hopkins University Press. 66, 83

Huff, Toby (2003). *The Rise of Early Modern Science. Islam, China, and the West. (second edition)*. Cambridge University Press. 29

Hughes, Thomas (1983). *Networks of Power: Electrification in Western Society, 1880–1930*. Baltimore: Johns Hopkins University Press. 22

Hughes, Thomas P. and Agatha C. Hughes, eds., (1990). *Lewis Mumford. Public Intellectual*. New York: Oxford University Press. 92

Illich, Ivan (1973). *Tools for Conviviality*. Harper and Row.

Irwin, Alan (1995). *Citizen Science*. Routledge. 136

Isaacson, Walter (2009). *Einstein. His Life and Universe*. London: Pocket Books. 81

Jacob, Margaret (1988). *The Cultural Meaning of the Scientific Revolution*. New York: Alfred Knopf. 30

Jamison, Andrew (1978). Democratizing Technology, in *Environment*, January-February. 136

Jamison, Andrew (1982). *National Components of Scientific Knowledge. A Contribution to the Social Theory of Science*. Lund: Research Policy Institute. 77

Jamison, Andrew (1988). Social Movements and the Politicization of Science, in J. Annerstedt and A. Jamison, eds., *From Research Policy to Social Intelligence*. Macmillan.

Jamison, Andrew (1994). Western science in perspective and the search for alternatives, in Jean-Jacques Salomon, et al., eds., *The Uncertain Quest. Science, Technology and Development*. Tokyo: The United Nations University Press. 57

Jamison, Andrew (1996). The Shaping of the Global Environmental Agenda: The Role of Non-Governmental Organizations, in Lash et al., eds., *Risk, Environment, Modernity*. Sage. 138

Jamison, Andrew (1998). American Anxieties: Technology and the Reshaping of Republican Values, in M Hård and A Jamison, eds., *The Intellectual Appropriation of Technology. Discourses on Modernity 1900–1939*. Cambridge, Ma: The MIT. Press 93

Jamison, Andrew (2001). *The Making of Green Knowledge. Environmental Politics and Cultural Transformation*. Cambridge University Press. 128, 131

Jamison, Andrew (2006). Social Movements and Science. Cultural Appropriations of Cognitive Praxis, in *Science as Culture*, March. DOI: 10.1080/09505430500529722 24

Jamison, Andrew (2008). To Foster a Hybrid Imagination: Science and the Humanities in a Commercial Age, in *NTM – Zeitschrift für Geschichte der Wissenschaften, Technik und Medizin*, No. 1.

Jamison, Andrew (2009). Educating Sustainable Architects. Reflections on the Alley Flat Initiative at the University of Texas. Unpublished manuscript. 143

Jamison, Andrew (2010a). In Search of Green Knowledge: A Cognitive Approach to Sustainable Development, in S Moore, ed. *Pragmatic Sustainability*. Routledge. 146

Jamison, Andrew (2010b). Climate Change Knowledge and Social Movement Theory, in *Wiley Interdisciplinary Reviews: Climate Change*, November. DOI: 10.1002/wcc.88 128, 142

Jamison, Andrew, Ron Eyerman, Jacqueline Cramer, and Jeppe Læssøe (1990). *The Making of the New Environmental Consciousness. A Comparative Study of the Environmental Movements in Sweden, Denmark, and the Netherlands*. Edinburgh University Press. 24, 128, 134

Jamison, Andrew and Ron Eyerman (1994). *Seeds of the Sixties*. University of California Press. 112, 115, 116

Jamison, Andrew and Mikael Hård (2003). The Story-Lines of Technological Change: Innovation, Construction, Innovation, in *Technology Analysis and Strategic Management*, Vol. 15, No. 1. DOI: 10.1080/0953732032000046060 14, 124

Jamison, Andrew and Niels Mejlgaard (2010). Contextualizing Nanotechnology Education: Fostering a Hybrid Imagination in Aalborg, Denmark, in *Science as Culture*, Vol. 19, No. 3. DOI: 10.1080/09505430903512911 18

Jardine, Lisa (1996). *Worldly Goods: A New History of the Renaissance*. London: Macmillan. 34

Jardine, Lisa (1999). *Ingenious Pursuits: Building the Scientific Revolution*. London: Little Brown. 47

Jennings, Humphrey (1987). *Pandaemonium. The Coming of the Machine as Seen by Contemporary Observers*. London: Picador. 56, 58

Jessen, Erland (2007). The romantic polytechnician in the philosophy and work of H. C. Ørsted, in Steen H. Christensen, et al., eds., *Philosophy in Engineering*. Aarhus: Academica. 62, 63

Johnson, Paul (2002). *The Renaissance. A Short History*. New York: Modern Library. 37

Kazin, Michael (1995). *The Populist Persuasion. An American History*. New York: Basic Books. 74

Kingston, Jos (1976). It's Been Said Before and Where Did That Get Us?, in Boyle and Harper, eds. 61

Klein, Naomi (2000). *No Logo: no space, no choice, no jobs*. Flamingo. 142

Krutch, Joseph Wood (1929). *The Modern Temper*. New York: Harcourt Brace Jovanovich.

Kuhn, Thomas (1962). *The Structure of Scientific Revolutions*. The University of Chicago Press. 113

Landes, David (1969). *The Unbound Prometheus. Technological Change and Industrial Development in Western Europe from 1750 to the Present*. Cambridge University Press. 29

Landes, David (1998). *The Wealth and Poverty of Nations. Why Some are so Rich and Some so Poor*. New York: Norton. 29

Latour, Bruno (1986). *Science in Action. How to Follow Scientists and Engineers through Society*. Cambridge, Ma: Harvard University Press. 21

Latour, Bruno (1993). *We Have Never Been Modern*. Cambridge, Ma: Harvard University Press. 2

Latour, Bruno (1996). *Aramis, or the Love of Technology*. Cambridge, Ma: Harvard University Press. 21, 107

Latour, Bruno (2005). *Reassembling the Social. An Introduction to Actor-Nertwork Theory*. Oxford University Press. 21, 108

Levy, Steven (1984). *Hackers. Heroes of the Computer Revolution*. New York: Dell. 120

Lilienthal, David (1966/1943). *TVA. Democracy on the March*. Chicago: Quadrangle Paperbacks. 95

Lilley, Samuel (1975). Technological Progress and the Industrial Revolution 1700–1914, in C Cipolla, ed. *The Fontana Economic History of Europe. Volume 3. The Industrial Revolution*. Glasgow: Fontana. 54, 55

Lovins, Amory (1977). *Soft Energy Paths*. Penguin.

Löwy, Michael and Robert Sayre (2001). *Romanticism against the tide of modernity*. Durham, NC: Duke University Press. 61

Lucena, Juan, Jen Schneider and Jon Leydens (2010). *Engineering and Sustainable Community Development*. Morgan & Claypool Publishers. 145

Lundvall, Bengt-Åke, ed. (1992). *National Systems of Innovation*. London: Pinter, 19, 67

MacCarthy, Fiona (1994). *William Morris. A Life for Our Times*. London: Faber and Faber. 71

Marcuse, Herbert (1964). *One-Dimensional Man*. Boston: Beacon.

Marinetti, Filippo (1960/1909). Manifesto of Futurism, reprinted in Eugen Weber, ed. *Paths to the Present: Aspects of European Thought from Romanticism to Existentialism*. New York. Dodd, Mead and Company. 87

Markoff, John (2005). *What the Dormouse Said. How the Sixties Counter-culture Shaped the Personal Computer Industry*. London: Penguin Books. 120

Marx, Karl (1973/1857). *The Grundrisse. Foundations of the Critique of Political Economy*. Translated by Martin Nicolaus. Harmondsworth: Penguin.

Marx, Karl (1976/1867). *Capital. A Critique of Political Economy. Volume 1*. Translated by Ben Fowkes. Harmondsworth: Penguin. 53

Marx, Karl and Frederick Engels (1968) *Selected Works*. Moscow: Progress Publishers. 53

Marx, Leo (1964), *The Machine in the Garden. Technology and the Pastoral Ideal in America*. Oxford University Press.

McKeon, Richard, ed. (1947). *Introduction to Aristotle. The essence of Aristotle's philosophy, including Organon, Physics, Ethics, On the Soul, Metaphysics, Politics and Poetics*. New York: The Modern Library. 5, 27

McNeely, Ian and Lisa Wolverton (2008). *Reinventing Knowledge. From Alexandria ti the Internet*. New York: Norton.

Merchant, Carolyn (1980). *The Death of Nature: Women, Ecology and the Scientific Revolution*. San Francisco: Harper & Row. 30

Merton, Robert (1942). Science and technology in a democratic society, in *Journal of Legal and Political Sociology*, 1. 10

Miller, Donald (1989). *Lewis Mumfod. A Life*. New York: Weidenfeld & Nicolson. 79, 93

Mitcham, Carl and David Muñoz (2010) *Humanitarian Engineering*. Morgan & Claypool. 114

Mokyr, Joel (1990). *The Lever of Riches. Technological Creativity and Economic Progress.* Oxford University Press. 29

Morris, William (1882). The Lesser Arts of Life. Accessed March 28, 2011 at `http://www.burrows.com/morris/lesser.html` 72

Morris, William (1993). *News From Nowhere and Other Writings.* London: Penguin. 72

Mumford, Lewis (1926). *The Golden Day. A Study in American Literature and Culture.* New York: Boni and Liveright.

Mumford, Lewis (1934). *Technics and Civilization.* New York: Harcourt Brace Jovanovich. 94

Mumford, Lewis (1961). Science as Technology, in *Proceedings of the American Philosophical Society,* Vol. 105, No. 5.

Mumford, Lewis (1970). *The Pentagon of Power.* New York: Harcourt Brace Jovanovich. 22, 46

Nandy, Ashis (1987). *Traditions, Tyranny and Utopias.* Delhi: Oxford University Press. 98

Nye, David (1994). *American Technological Sublime.* Cambridge Ma: The MIT Press. 87

Nye, David (2006). *Technology Matters.* Cambridge, MA: The MIT Press. 23

Ovitt, George (1987). *The Restoration of Perfection: Labour and Technology in Medieval Culture.* New Brunswick, NJ: Rutgers University Press. 32, 33

Pacey, Arnold (1974). *The Maze of Ingenuity. Ideas and Idealism in the Development of Technology.* London: Allen Lane. 32

Pachter, Henry (1961). *Paracelsus: Magic into Science.* New York: Collier Books. 39, 40

Parks, Bradley and J. Timmons Roberts (2010). Climate Change, Social Theory and Justice, in *Theory, Culture and Society* Vol. 27, No. 2-3. DOI: 10.1177/0263276409359018 143

Price, Derek de Solla (1963). *Little Science, Big Science.* Columbia University Press. 109

Reader, John (2006). *Globalization, Engineering and Creativity.* Morgan & Claypool. 1

Reich, Charles (1970). Reflections: The Greening of America, in *New Yorker,* Sept 26. 128

Roszak, Theodore (1969). *The Making of a Counter Culture.* Garden City, NY: Anchor Books. 119, 120

Rowell, Andrew (1996). *Green Backlash. Global Subversion of the Environment Movement.* Routledge. 137

Rushdie, Salman (1992). *Imaginary Homelands. Essays and Criticism 1981–1991*. London: Granta Books. 2

Schlosberg, David (2007). *Defining Environmental Justice: Theories, Movements and Nature*. Oxford University Press. 143

Schumpeter, Joseph (1975/1942). *Capitalism, Socialism and Democracy*. New York: Harper and Row. 19

Sen, Amartya (2005). *The Argumentative Indian. Writings on Indian History, Culture and Identity*. New York: Farrar, Straus and Giroux. 85, 97

Shapin, Steven (1996). *The Scientific Revolution*. The University of Chicago Press. 29

Shapin, Steven (2008). *The Scientific Life. A Moral History of a Late Modern Vocation*. The University of Chicago Press. 2, 29

Shelley, Mary (1994/1818). *Frankenstein, or The Modern Prometheus. The 1818 Text*. Oxford University Press. 59

Shiva, Vandana (1988). *Staying Alive. Women, Ecology and Development*. Zed Books. 147

Shiva, Vandana (2000). *Stolen Harvest. The Hijacking of the Global Food Supply*. South End Press. 147

Shiva, Vandana (2005). *Earth Democracy. Justice, Sustainability and Peace*. South End Press. 147

Simon, Herbert (1969). *Sciences of the Artificial*. The MIT Press. 124

Slaughter, Sheila and Gary Rhoades (2004). *Academic Capitalism and the New Economy*. Baltimore: Johns Hopkins University Press. 2

Suzuki, David and Dave Robert Taylor (2009). *The Big Picture. Reflections on science, humanity, and a quickly changing planet*. Vancouver: Greystone Books.

Taylor, Bron, ed. (1995). *Ecological Resistance Movements. The Global Emergence of Radical and Popular Environmentalism*. Albany: State University of New York Press. 143

Thoreau, Henry David (1842). Natural History of Massachusetts. Accessed March 29, 2011 at `http://www.thoreau-online.org/natural-history-of-massachusetts.html`

Tobin, James (2003). *First to Fly. The unlikely triumph of Wilbur and Orville Wright*. London: John Murray. 75, 76

Turner, Fred (2006). *From Counterculture to Cyberculture. Stewart Brand, the Whole Earth Network, and the Rise of Digital Utopianism*. The University of Chicago Press. 121

Uglow, Jenny (2002). *The Lunar Men. Five Friends Whose Curiosity Changed the World.* New York: Farrar, Straus and Giroux. 49

Visvanathan, Shiv (1984). *Organizing for Science. The Making of an Industrial Research Laboratory.* Delhi: Oxford University Press. 99

von Wright, Georg Henrik (1978). *Humanismen som livshållning.* Stockholm: MånPocket. 3

Walls, Laura (1995). *Seeing New Worlds: Henry David Thoreau and nineteenth-century natural science.* Madison: University of Wisconsin Press. 66

Ward, Barbara and René Dubos (1972). *Only One Earth.* Penguin. 134

Weber, Max (1958). *From Max Weber: Essays in Sociology,* edited by H.H. Gerth and C. Wright Mills. Oxford University Press. 84

Weber, Max (2001/1904). *The Protestant Ethic and the Spirit of Capitalism.* London: Routledge. 20, 29

Webster, Charles (1975). *The Great Instauration: Science, Medicine and Reform 1626–1660.* New York: Holmes and Meier. 30, 46

Weinberg, Alvin (1967). *Reflections on Big Science.* The MIT Press. 111

White, Lynn (1962). *Medieval technology and social change.* Oxford University Press. 29

White, Lynn (1978). *Medieval Religion and Technology. Collected Essays.* Berkeley: The University of California Press. 32, 34

Whitehead, Alfred North (1925). *Science and the Modern World.* New York: Macmillan. 54, 60

Whitehead, Alfred North (1929). *The Aims of Education.* New York: Mentor.

Whitford, Frank (1984). *Bauhaus.* London: Thames & Hudson. 90

Williams, Raymond (1958). *Culture and Society 1780–1950.* Harmondsworth: Penguin. 22

Williams, Raymond (1977). *Marxism and Literature.* Oxford University Press. 25, 52

Williams, Raymond (1989). *Politics of Modernism. Against the New Conformists.* London: Verso. 87

Winstanley, Gerrard (1973/1652). *The Law of Freedom and Other Writings.* Harmondsworth: Penguin.

Worldwatch Institute (2010). *State of the World 2010.* London: Earthscan. 143

Worster, Donald (1979). *Nature's Economy. The Roots of Ecology.* Garden City, NY: Anchor Books. 129

162   BIBLIOGRAPHY

Yates, Frances (2002/1964). *Giordano Bruno and the Hermetic Revolution*. London: Routledge. 30

Yates, Frances (1972). *The Rosicrucian Enlightenment*. London: Routledge & Kegan Paul. 36, 43

Yoxen, Edward (1983). *The Gene Business. Who should control Biotechnology?* Pan Books. 123

Zilsel, Edgar (2000). *The Social Origins of Modern Science*. Dordrecht: Kluwer. 34

Authors' Biographies

The authors have written this book as part of their work together in the Program of Research on Opportunities and Challenges in Engineering Education in Denmark (PROCEED), which is supported by the Danish Strategic Research Council.

ANDREW JAMISON

Andrew Jamison is professor of technology, environment and society at the Department of Development and Planning at Aalborg University and coordinator of PROCEED. He has a BA in history and science from Harvard University (*magna cum laude,* 1970) and a PhD in theory of science from Gothenburg University (1983). He has been teaching natural science and engineering students about the social and cultural contexts of science and technology since the early 1970s. Professor Jamison has published widely in the fields of science and technology policy and environmental politics and is the author, most recently, of *The Making of Green Knowledge. Environmental Politics and Cultural Transformation* (Cambridge University Press 2001) and, with Mikael Hård, *Hubris and Hybrids. A Cultural History of Science and Technology* (Routledge 2005).

For more information, including downloadable lectures and articles, and even some songs, please visit Professor Jamison's website: `http://people.plan.aau.dk/~andy/`

STEEN HYLDGAARD CHRISTENSEN

Steen Hyldgaard Christensen is senior lecturer at the Institute of Business and Technology in Herning, which is a part of Aarhus University, where he teaches literature and the history of ideas, research methodology and philosophy of science. He has an MA in Scandinavian Language and Literature and the History of Ideas, from Aarhus University. He has been the initiator and coordinator of three projects on engineering and culture, and he is the co-editor of *Philosophy in Engineering* (Academica 2005) and *Engineering in Context* (Academica 2009). He was the initiator of PROCEED and he is the editor-in chief of the forthcoming book, co-sponsored by PROCEED, *Engineering, Philosophy and Development: American, Chinese and European Perspectives*, to be published by Springer Verlag.

LARS BOTIN

Lars Botin is an assistant professor at the Department of Development and Planning at Aalborg University, where he teaches theory of science and contextual knowledge to engineering students, and in a number of other departments, as well. He has an MA in art history from Aarhus University and a PhD from Aalborg University (2008), with a dissertation entitled, *A Humanist in the Hospital:*

Cultural Assessment of Electronic Health Records, in which he presents a phenomenological approach to ethnographic research, using a video observation methodology.

Lightning Source UK Ltd.
Milton Keynes UK

177998UK00003B/86/P